Nano Plastics

Nano Plastics

The Invisible Threat Inside Us

Robert Sudar

Published by RCS Experience Sheridan, WY https://rcs-experience.com/books

ISBNs Paperback: 978-1-971877-04-4

Library of Congress Control Number: 2026905437
U.S. Copyright Office Registration: TXu 2-564-308
Effective Date: February 25, 2026

Ordering Information: Quantity sales. Special discounts are available on quantity purchases by corporations, associations, and others. For details, contact the publisher at the address above.

Disclaimers: The information in this book is for educational and informational purposes only and is not intended as medical, legal, or professional advice. While every effort has been made to ensure the accuracy of the information herein, the author and publisher assume no responsibility for errors or omissions. Always seek the advice of a qualified professional regarding personal health or environmental safety.

Printed in the United States of America. First Edition: 2026

To the memory of Doug Timm,
whose brief but meaningful friendship inspired me.

His enthusiasm for music was contagious.

It was an honor to know him and work with him.

Here's to you "Mad-Dog."

Introduction

There is a form of pollution you cannot see.

It does not look like litter.
It does not float in obvious patches.
It does not announce itself.

But it is there—around you, within your environment, and increasingly, within the human body.

This is the world of nanoplastics.

Plastic has become one of the defining materials of modern life.

It makes products lighter, cheaper, and more durable.
It protects food.
It enables technology.
It supports medicine and infrastructure.

In many ways, it has made life more convenient and more efficient.

But plastic does not disappear.

Over time, it breaks down—first into microplastics, and eventually into particles so small they can no longer be seen without specialized equipment.

These are nanoplastics.

At this scale, the conversation changes.

We are no longer just talking about environmental pollution. We are talking about particles that can:

- move through air and water
- enter the human body through everyday exposure
- and interact with biological systems in ways science is only beginning to understand

This is where concern begins—not from speculation, but from emerging evidence.

This book is not about fear.

It is about understanding.

You will not find exaggerated claims or dramatic conclusions here. What you will find is a clear, grounded look at:

- what nanoplastics are
- how they form
- how they enter the body
- what current research suggests about their behavior
- and what remains uncertain

Because both sides matter.

Knowing what we **do know** helps us take this seriously.
Knowing what we **don't know** keeps us honest.

You will also find something else:

Practical guidance.

Not extreme solutions.
Not unrealistic expectations.
But realistic ways to:

- reduce exposure
- make better everyday decisions
- and support your body in a world where plastic is unavoidable

This is not about eliminating plastic from your life.

That is not realistic.

It is about changing how you interact with it.

There is a simple truth behind everything in this book:

> Plastic, in any form, is not something the human body evolved to contain.

Whether it remains relatively inert or actively interacts with cells, its presence represents a form of exposure that did not exist for most of human history.

Understanding that is not about fear.

It is about perspective.

We are living through a shift in awareness.

What was once considered harmless is now being examined more closely.
What was once invisible is now being measured.

And with that comes a responsibility:

To understand the world as it is, not as we assumed it to be.

This book is your guide to that understanding.

Not to alarm you.

But to inform you.

So that you can move forward—not in fear—but with clarity, awareness, and control over the choices that matter most.

Table of Contents

PART 1
A Problem You Can't See

CHAPTER 1

The Smallest Pollution Crisis in History

There is a form of pollution that you cannot see, cannot taste, and cannot avoid simply by stepping outside or staying in. It doesn't sit on the surface of oceans or collect in visible piles. It drifts through the air in your home, settles into your food, mixes into your drinking water, and—most concerning of all—enters your body.

This is the world of **nanoplastics**.

Most people have heard of plastic pollution: bottles in the ocean, bags in landfills, debris washing up on beaches. Over time, those larger plastics break down into smaller fragments known as **microplastics**—tiny particles smaller than a grain of rice. That alone is a serious environmental problem.

But the story doesn't end there.

As plastics continue to degrade, they become even smaller—so small that they cross into a scale measured in **nanometers**. These are **nanoplastics**, and they represent a different kind of risk entirely.

What Are Nanoplastics? (A Simple Explanation)

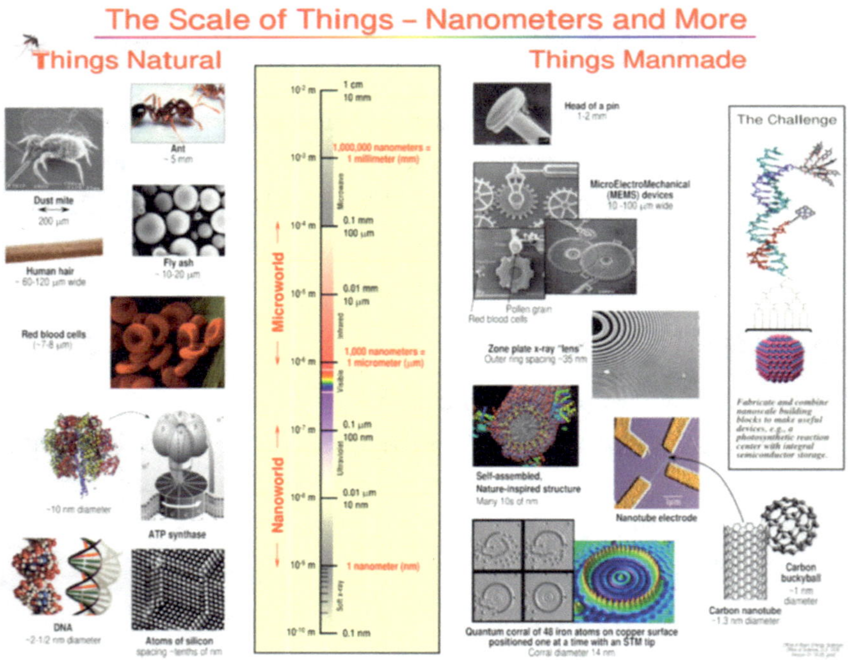

"The Scale of Things - Nanometers and More" (U.S. Department of Energy / NIST). Public domain

To understand nanoplastics, you need a sense of scale.

A **nanometer** is one-billionth of a meter. That number is almost impossible to visualize, so let's bring it into perspective:

- A human hair is about **80,000–100,000 nanometers wide**
- A red blood cell is about **7,000 nanometers across**
- Many nanoplastics are **under 1,000 nanometers**, and some are far smaller

At this scale, plastic particles are no longer just tiny pieces of debris. They behave differently. They can move more freely, interact more intimately with biological systems, and reach places in the body that larger particles cannot.

Nanoplastics are not a specific type of plastic—they are defined by **size**. They can originate from common materials like:

- Polyethylene (used in bags and packaging)
- Polystyrene (used in foam and containers)
- Polypropylene (used in food containers and textiles)

What changes is not the material—but the **scale at which it exists**.

Microplastics vs. Nanoplastics: Why Size Changes Everything

It's tempting to think of nanoplastics as simply "smaller microplastics." But that assumption misses the point.

Size is not just a detail—it changes behavior.

Here's a useful way to think about it:

- **Microplastics** are like **sand** You can see them. They may irritate or accumulate, but they largely stay outside cells.
- **Nanoplastics** are more like **smoke** They move easily, spread widely, and can penetrate deep into spaces that are otherwise protected.

A

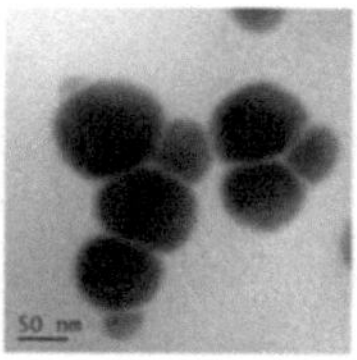

Adapted from Hao et al., "Exposure to polystyrene nanoparticles..." PLOS ONE (2025), Fig 1A. Licensed under CC BY

Because of their size, nanoplastics can:

- Pass through cell membranes
- Enter the bloodstream
- Interact with proteins and cellular structures
- Potentially cross protective barriers in the body

This is where concern begins to grow.

Why Smaller Can Mean More Dangerous

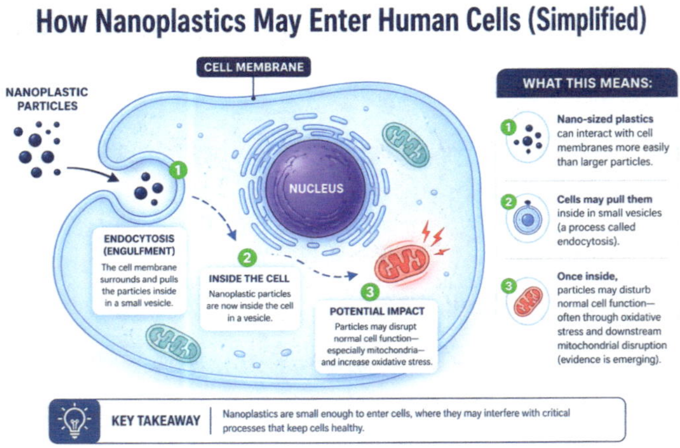

Larger plastic particles tend to stay in the digestive system or become trapped in tissues. Nanoplastics, on the other hand, operate on a different level.

Emerging research suggests that nanoplastics may be able to:

- **Enter individual cells**
- **Interact with cellular components**, including mitochondria

- **Trigger oxidative stress and inflammation**
- **Cross biological barriers**, such as the gut lining and possibly the blood-brain barrier

This does not mean we fully understand the extent of harm. In fact, one of the most important things to acknowledge is this:

> **The science is still developing.**

What we do have is a growing body of evidence pointing to **biological interaction**, not just passive presence.

That distinction matters.

It shifts the conversation from:

> "Plastic is around us"

to:

> **"Plastic may be interacting with us."**

Where Nanoplastics Come From

Nanoplastics are not manufactured intentionally in most cases. They are the result of **breakdown**.

Everyday plastic items degrade over time due to:

- Sunlight (UV radiation)
- Heat
- Mechanical stress (friction, washing, wear)
- Chemical exposure

As they break apart, they move from visible fragments to microscopic pieces—and eventually to nanoscale particles.

This process is happening constantly, all around us.

1. Air (One of the Most Overlooked Sources)

Indoor air can contain plastic particles from:

- Synthetic fabrics (polyester, nylon)
- Carpets and upholstery
- Household dust

Each time fabrics are worn, washed, or disturbed, they can release microscopic fibers that continue to break down.

2. Water

Nanoplastics have been detected in:

- Bottled water
- Tap water
- Oceans, lakes, and rivers

Water treatment systems were never designed to remove particles at the nanoscale.

3. Food

Food can become contaminated through:

- Packaging materials
- Processing environments
- Environmental exposure (soil, water, air)

Heating food in plastic containers can accelerate particle release.

4. Clothing and Textiles

Synthetic clothing sheds microfibers continuously. Over time, these fibers fragment further into nanoplastics.

5. Everyday Packaging

Plastic wraps, containers, and coatings degrade—especially when exposed to heat, friction, or repeated use.

An Invisible, Continuous Exposure

Unlike many environmental risks, nanoplastics are not tied to a single source or event.

You are not exposed once.

You are exposed **continuously**.

- When you breathe indoor air
- When you drink water
- When you eat food
- When you wear clothing

This doesn't mean panic is the right response. But it does mean awareness is necessary.

What We Know—and What We Don't

At this stage, the science is clear on a few key points:

- Nanoplastics exist in the environment and in human exposure pathways
- Their small size allows them to interact with biological systems in ways larger particles cannot
- Laboratory studies show they can affect cells under certain conditions

What remains uncertain is just as important:

- How much accumulates in the human body over time
- What levels of exposure are harmful
- How the body processes or eliminates them
- The long-term health effects of chronic, low-level exposure

This is why many researchers describe nanoplastics not as a confirmed crisis—but as an **emerging concern with serious implications**.

A Different Kind of Pollution

Most pollution we worry about is visible, measurable, and often avoidable with clear steps.

Nanoplastics are different.

They are:

- **Invisible**
- **Ubiquitous**

- **Persistent**
- **Biologically active (based on emerging evidence)**

And perhaps most importantly:

> They are already part of our environment—and likely part of us.

Why This Matters Now

For decades, plastic was seen as a convenience. Durable, flexible, and cheap, it became embedded in nearly every part of modern life.

Only recently have we begun to understand the consequences of that durability.

Plastic does not disappear.

It transforms.

And at its smallest scale, it may begin to interact with the very systems that keep us alive.

This book is not about fear.

It is about understanding:

- what is happening,
- what the science suggests,
- and what can realistically be done about it.

Because the smallest pollution crisis in history is not somewhere far away.

It is already here.

CHAPTER 2

How They Get Inside You (And Stay There)

If nanoplastics were rare, or limited to specific environments, this would be a much simpler story.

But they're not.

They are present in the air you breathe, the water you drink, the food you eat, and the materials you interact with every day. And because of their size, they don't just pass by you—they can enter your body.

Not once. Not occasionally.

Continuously.

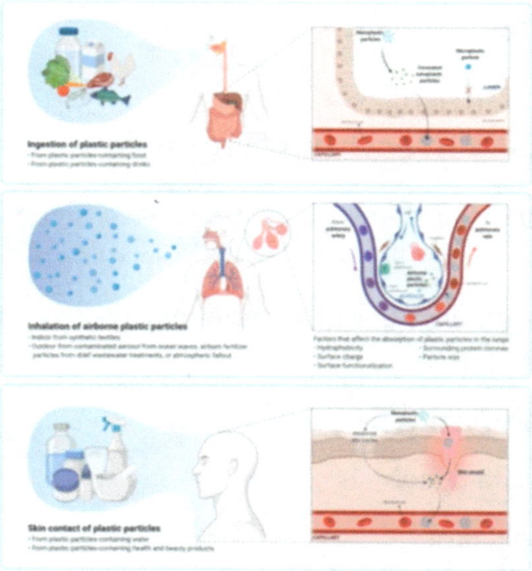

Reproduced from Impact of Microplastics and Nanoplastics on Human Health, Nanomaterials (2021), Figure 2, under CC BY 4.0 (MDPI)

You Don't Notice the Entry Point

Unlike many environmental exposures, there is no single moment where nanoplastics "enter" your body. There's no smell, no taste, no immediate reaction.

Exposure happens quietly, through normal, everyday activities:

- breathing
- eating
- drinking
- simply existing in modern indoor environments

And while the science is still developing, one thing is clear:

> **Human exposure is not hypothetical—it is already happening.**

Inhalation — Breathing Them In

One of the most overlooked pathways is also one of the most constant: **air**.

Indoor air, in particular, can contain microscopic plastic particles originating from:

- synthetic clothing (polyester, nylon)
- carpets and upholstery
- household dust
- furniture and insulation materials

Every time fabrics are worn, washed, or disturbed, they shed tiny fibers. Over time, these fibers break down into smaller and smaller fragments—eventually reaching the nanoscale.

When you breathe, these particles can be drawn into the respiratory system.

What happens next is still being studied, but early research suggests:

- some particles may become trapped in the upper airways
- smaller particles may reach deeper into the lungs
- nanoscale particles may interact with lung tissue at a cellular level

This doesn't mean every particle causes harm. But it does mean the lungs are a **direct interface between the environment and the bloodstream**—and nanoplastics are small enough to take advantage of that.

Ingestion — Eating and Drinking Them

Nanoplastics have been detected in:

- drinking water (both bottled and tap)
- seafood and processed foods
- food exposed to plastic packaging

The digestive system is another major entry point.

Particles can enter the body through:

- contaminated water
- food processing environments

- packaging materials, especially when heated

Inside the gut, the body is designed to absorb nutrients—but that same system can also allow very small particles to pass through the intestinal lining.

Emerging research suggests:

- some nanoplastics may cross the **gut barrier**
- they may enter circulation or interact with gut cells
- they may influence inflammation or microbiome balance (still under investigation)

Again, the key point isn't that every exposure leads to damage—it's that **entry is possible**, and likely occurring.

Skin Exposure — Minor, But Not Zero

Compared to inhalation and ingestion, the skin is a more effective barrier.

But it's not perfect.

Nanoplastics may come into contact with skin through:

- personal care products
- synthetic clothing
- environmental exposure (dust, surfaces)

Under normal conditions, most particles are unlikely to penetrate deeply. However:

- very small particles
- damaged or compromised skin

- prolonged exposure

may increase the chance of interaction.

This pathway is considered **less significant**, but not irrelevant—especially as research continues.

Crossing the Body's Protective Barriers

What makes nanoplastics different is not just how they enter the body—but where they may go next.

The human body has evolved protective barriers to keep harmful substances out of critical systems. These include:

The Gut Barrier

The intestinal lining is designed to allow nutrients in while keeping unwanted materials out.

But nanoscale particles are small enough to:

- pass between cells (in some conditions)
- be taken up by specialized cells
- interact directly with intestinal tissue

This is one of the most studied entry points—and one of the clearest examples of how size changes behavior.

The Blood–Brain Barrier

The brain is protected by a highly selective filter known as the **blood–brain barrier**.

Its job is to keep most substances out of the brain's environment.

Emerging research suggests that very small particles, including some nanoparticles, may:

- cross or bypass this barrier under certain conditions
- interact with neural tissue

For nanoplastics specifically, this area is still under active investigation. Some studies indicate the possibility of transport into brain tissue, but the extent and impact are not yet fully understood.

> This is a critical area of concern—but also one where the science is still evolving.

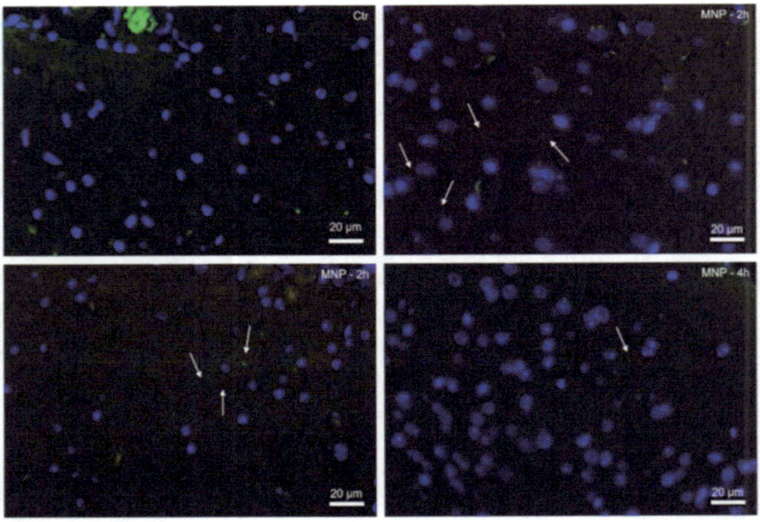

Reproduced from Kopatz et al., Micro- and Nanoplastics Breach the Blood-Brain Barrier (BBB): Biomolecular Corona's Role Revealed, Nanomaterials (2023), Figure 3, under CC BY 4.0 (MDPI)

The Placenta

During pregnancy, the placenta acts as a protective interface between mother and developing fetus.

Recent studies have detected microscopic plastic particles in placental tissue, raising important questions about:

- maternal exposure
- transfer during development
- long-term effects on early life

Whether nanoplastics behave the same way as larger particles—and to what extent—is still being studied. But the implication is clear:

> These particles are capable of reaching some of the most protected environments in the human body.

Why Size Changes Everything (Again)

At larger scales, particles are easier for the body to isolate, expel, or ignore.

At the nanoscale:

- particles can move more freely
- interact more closely with cells
- potentially cross biological barriers

This doesn't automatically make them dangerous—but it does make them **biologically relevant**.

And that's the shift happening in current research.

Do They Stay in the Body?

This is one of the most important—and still evolving—questions in nanoplastics research.

Early studies have already detected plastic particles in human tissues, including the lungs, blood, placenta, and in some cases, brain tissue. These findings suggest that at least some particles are not immediately cleared by the body.

In studies examining post-mortem tissue, researchers have reported the presence of plastic particles in the brain, raising new questions about long-term accumulation. Some findings have also suggested higher concentrations in individuals with neurodegenerative conditions.

However, these studies are still limited in size and scope, and methods for detecting and measuring nanoplastics are still being refined.

What remains unclear is:

- how much accumulates over a lifetime
- how long these particles persist in different tissues
- and whether they play a direct role in disease, or are simply present alongside it

At the same time, researchers are investigating how the body may remove these particles, including:

- processing through the liver
- filtration by the kidneys
- elimination through bile or waste

The picture is still developing.

But the combination of:

- continuous exposure
- the ability of particles to enter the body
- and early evidence of their presence in human tissues

is what is driving increasing scientific and public concern.

The Reality of Daily Exposure

You are not exposed once.

You are exposed:

- when you breathe indoor air
- when you drink water
- when you eat packaged or processed food
- when you interact with synthetic materials

And because nanoplastics are so small, these exposures don't require unusual conditions.

They happen during normal life.

Plastic, regardless of size, is not a natural component of the human body.

Whether it remains relatively inert or actively interacts with cells, its presence still represents exposure to a foreign material—one the body did not evolve to process or accommodate.

What We Know — and What We're Still Learning

What we know:

- Nanoplastics are present in air, water, and food
- Multiple entry pathways exist (inhalation, ingestion, skin)
- Their size allows interaction with biological systems

What we don't fully know:

- How much enters and remains in the body
- How efficiently the body clears them
- The long-term effects of chronic, low-level exposure

The Takeaway

This is not about a single exposure event or a rare condition.

It is about **constant, low-level contact** with particles that are small enough to interact with the body in new ways.

And while the full picture is still coming into focus, one conclusion is already clear:

You are already exposed—every day.

The question now is not whether nanoplastics enter the body.

The question is:

What happens after they do?

PART 2
What Happens Inside the Body

CHAPTER 3

The Role of Surface Charge (Simple but Powerful)

If size explains how nanoplastics can enter the body, **surface charge helps explain what they do once they're inside**.

This is one of the most important—and often misunderstood—parts of the story.

It's also where things can easily be overstated.

So let's keep it simple, accurate, and grounded in what science actually supports.

What Is "Surface Charge"? (In Plain English)

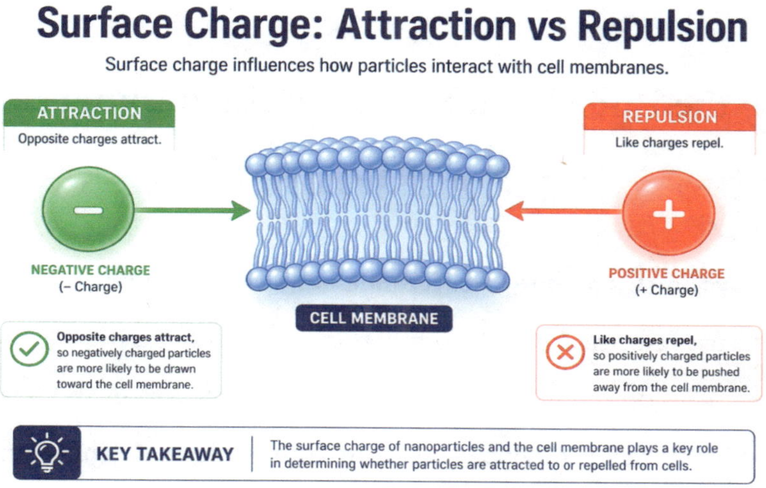

You've experienced surface charge before, even if you didn't call it that.

Think about:

- rubbing a balloon on your hair
- clothes sticking together in a dryer
- static shock when touching a metal handle

That's **static electricity**—a buildup of electrical charge on a surface.

Now imagine something similar happening at a scale so small you can't see it.

Nanoplastics can carry **tiny electrical charges on their surfaces**. Scientists measure this as something called **zeta potential**, but you don't need the term to understand the idea.

What matters is this:

> **Some particles tend to attract. Some tend to repel. And that affects how they behave around cells.**

Why the Body Is an Electrical Environment

Your body is not just chemical—it's also electrical.

Every cell:

- maintains a **charge difference across its membrane**
- uses electrical signals to function
- relies on charged molecules to move substances in and out

In the brain, this is obvious—neurons communicate through electrical impulses.

But even outside the brain, electrical properties matter:

- cell membranes have charged surfaces
- proteins carry electrical charges
- ions (like calcium and sodium) move based on electrical gradients

So when a charged particle enters this environment, it doesn't just drift passively.

It **interacts.**

How Charge Influences What Nanoplastics Do

Surface charge doesn't act alone—but it plays a key role in shaping how nanoplastics behave.

Think of it as a kind of "interaction setting" that affects what happens next.

1. Sticking to Cells

Cell membranes are not neutral. They carry their own electrical properties.

Because of this:

- some particles are more likely to **adhere to the surface of a cell**
- others may be **repelled or less interactive**

Laboratory studies have shown that:

- certain charged nanoparticles bind more strongly to cell membranes

- this can increase the chance of further interaction

In simple terms:

> **Charge influences whether a particle stays near a cell —or moves on.**

2. Entering Cells

Getting close to a cell is only the first step.

For a particle to have deeper effects, it often needs to be taken inside.

Cells can internalize particles through processes like:

- membrane engulfment (endocytosis)
- transport through specialized pathways

Surface charge can influence:

- how easily a particle is recognized and absorbed
- how strongly it interacts with the membrane during this process

Some types of charged particles are taken up more readily in experimental models.

That doesn't mean all nanoplastics behave the same way—but it does show that:

> **Charge can affect how likely a particle is to cross from "outside" to "inside."**

3. Interacting With Proteins

Once inside the body, nanoplastics don't remain "bare."

Almost immediately, they become coated with:

- proteins

- lipids
- other biological molecules

This coating is sometimes called a **"protein corona."**

And here's where things get interesting:

- the original surface charge helps determine **which molecules attach**
- this new outer layer then determines how the particle behaves inside the body

So even if the charge changes after entering the body, it still plays a role in shaping what happens next.

Charge influences the first interaction—which can shape everything that follows.

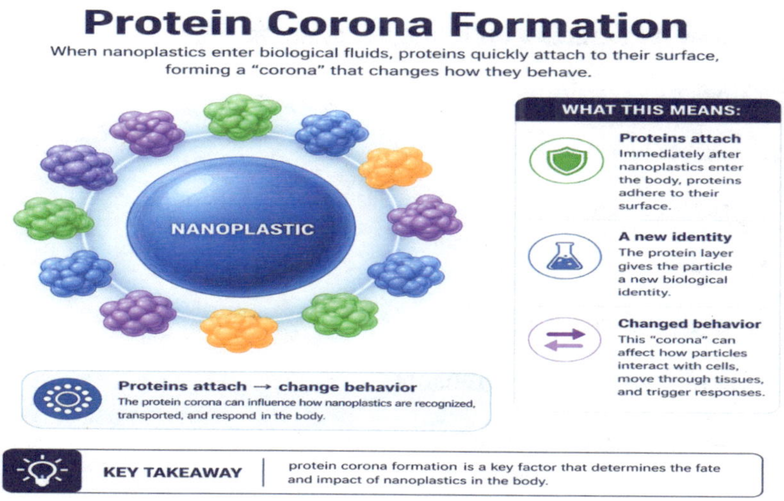

Why This Matters—Without Overstating It

It would be easy to say:

> "Surface charge is the main reason nanoplastics are dangerous."

But that would be too simplistic—and not supported by current science.

Instead, a more accurate way to understand it is:

> **Surface charge helps determine how aggressively a nanoplastic particle interacts with the body.**

It affects:

- how strongly it sticks
- how easily it enters cells
- how it interacts with biological systems

But it does not act alone.

Other factors also matter:

- particle size
- shape
- chemical composition
- additives and contaminants
- the biological environment it enters

What Happens After Entry

Once a nanoplastic particle is inside a cell, the story becomes more complex.

Researchers are studying how these particles may:

- trigger **oxidative stress**
- interfere with normal cellular processes
- affect structures like mitochondria

Surface charge plays a role in getting the particle to that point—but the downstream effects involve multiple biological pathways.

What We Know—and What We're Still Learning

What we know:

- Nanoparticles can carry surface charge
- Charge influences how particles interact with cells and proteins
- These interactions can affect uptake and biological response

What we don't fully know:

- how different types of nanoplastics behave in real human conditions
- how surface charge changes once particles are inside the body
- how much charge contributes to long-term health effects compared to other factors

The Takeaway

Surface charge is not the whole story.

But it is a critical part of it.

It helps explain why nanoplastics are not just tiny pieces of debris—but **active participants** in the biological environment they enter.

And while the science is still developing, one idea is becoming increasingly clear:

> **At the nanoscale, even something as subtle as electrical charge can shape how a particle interacts with the human body.**

Not as the sole cause of harm—but as one of the factors that determines how far that interaction goes.

CHAPTER 4

Breaking Into Cells

If nanoplastics can reach the body, and surface charge influences how they behave, the next question is unavoidable:

How do they actually get inside cells?

This is where the story becomes more biological—and more important.

Because once a particle enters a cell, it is no longer just present.

It becomes part of the system.

Attaching to the Cell Surface

Before anything enters a cell, it has to interact with the outer boundary—the **cell membrane**.

This membrane is not a solid wall. It is a dynamic, flexible structure made of lipids and proteins, constantly interacting with its environment.

Nanoplastics approaching a cell may:

- drift past without interaction
- loosely contact the surface
- or **bind more strongly**, depending on their properties

One of the key factors influencing this is **surface charge**.

Because cell membranes carry their own electrical properties:

- some particles are more likely to **stick**

- others are less likely to remain in contact

This initial attachment matters.

> If a particle does not stay near the membrane, it is far less likely to enter the cell.

Endocytosis — How Cells Pull Things Inside

Cells are not passive. They actively take in materials from their surroundings through a process called **endocytosis.**

In simple terms, this means:

- the cell membrane wraps around a particle
- forms a small pocket
- and pulls it inside

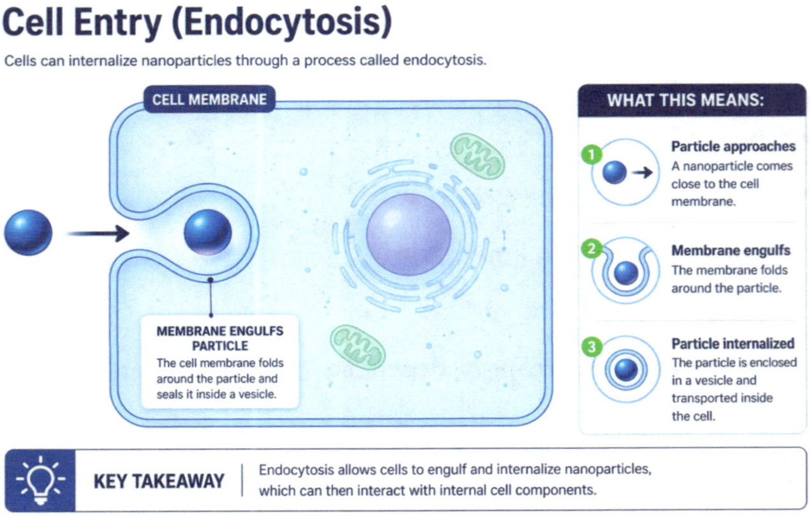

You can think of it as the cell briefly "swallowing" something from its environment.

This process is normal and necessary. Cells use it to:

- absorb nutrients
- regulate signals
- remove debris

But nanoplastics can take advantage of this same system.

Once attached to the membrane, a particle may trigger or be included in this engulfment process.

> The cell does not "recognize" plastic as plastic—it responds to size, shape, and surface properties.

Why Some Particles Enter More Easily

Not all nanoplastics behave the same way.

Laboratory studies show that some particles are taken up more readily than others.

One consistent pattern:

Positively charged particles tend to interact more strongly with cell membranes.

Why?

Because:

- cell membranes often carry a **net negative charge**
- opposite charges attract

This can lead to:

- stronger attachment
- increased likelihood of uptake
- more frequent internalization in experimental models

That doesn't mean every positively charged particle is harmful.

But it does mean:

> **Surface charge can influence how easily a particle crosses from outside the cell to inside.**

This is one of the reasons charge is described as affecting how "aggressive" a particle may be in biological systems.

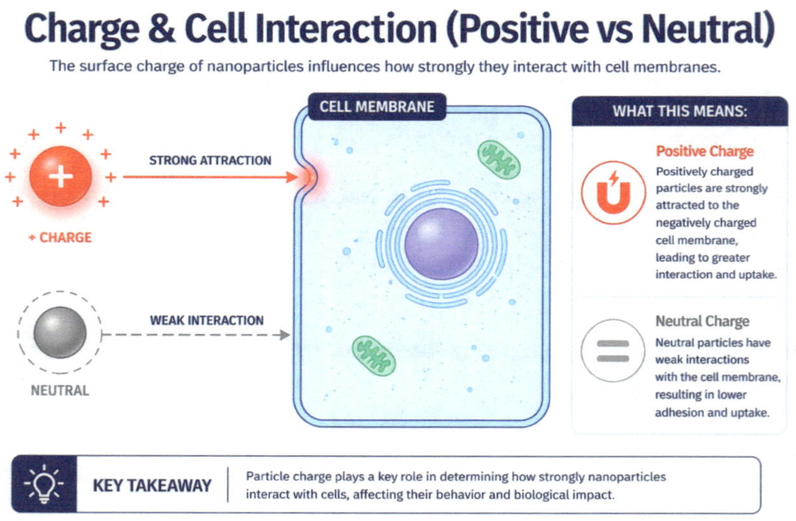

The Protein Corona — A Disguise in Plain Sight

Once nanoplastics enter the body, they don't remain in their original form for long.

Almost immediately, they become coated with biological molecules:

- proteins
- lipids
- other compounds found in blood and tissues

This layer is called the **protein corona**.

And it changes everything.

Instead of interacting with cells as "bare plastic," the particle now presents a new surface—one made of the body's own molecules.

You can think of it like this:

> The particle puts on a disguise.

This disguise can:

- alter how the immune system sees the particle
- change how cells respond to it
- influence where it travels in the body

In some cases, it may even make the particle more likely to be taken up by cells.

So while surface charge helps determine initial interactions, the protein corona reshapes what happens next.

What Happens After Entry

Once inside a cell, nanoplastics are typically enclosed in small compartments known as **vesicles**.

From there, several outcomes are possible:

- they may remain contained
- they may interact with cellular structures
- they may contribute to stress responses

Researchers are actively studying how these particles may:

- influence oxidative stress
- interfere with normal cell function
- affect energy production systems like mitochondria

This is where the science becomes more complex—and where many questions remain.

What We Know—and What We're Still Learning

What we know:

- Nanoplastics can attach to cell membranes
- Cells can internalize nanoscale particles through endocytosis
- Surface charge influences how strongly particles interact and enter
- Protein coronas form quickly and alter particle behavior

What we don't fully know:

- how often these processes occur in real human exposure conditions
- how different types of nanoplastics behave inside different cell types
- the long-term consequences of repeated cellular uptake

The Takeaway

Cells are designed to interact with their environment.

Nanoplastics, because of their size and surface properties, can take advantage of those same systems.

They can:

- attach
- be taken in
- and become part of the cellular environment

Not because the body is designed to accept them—

but because the mechanisms that keep us alive can also allow them in.

Chapter 5

The Mitochondria Problem (Energy Collapse)

If nanoplastics can enter cells, the next question is more serious:

What happens when they get inside?

To understand that, you need to understand one of the most important parts of every cell in your body:

the mitochondria.

These tiny structures are often described as the cell's **"power plants"** or **"batteries."**

And that description is not an exaggeration.

What Mitochondria Do (A Cell's Power Supply)

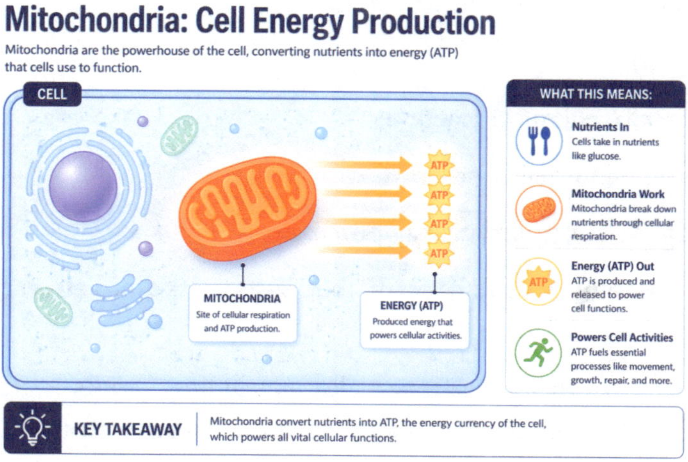

Every function in your body requires energy.

- your heart beating
- your brain thinking
- your muscles moving
- your cells repairing themselves

That energy is produced inside mitochondria.

They convert nutrients from food into a usable form of energy called **ATP (adenosine triphosphate)**.

You don't need to remember the name. Just remember this:

Without mitochondria, cells cannot function.

And when mitochondria don't work properly, everything begins to slow down—or break down.

What Happens When Mitochondria Are Disrupted

Mitochondria are sensitive structures.

They rely on:

- precise chemical reactions
- stable internal conditions
- tightly controlled energy processes

When something interferes with that system, the effects can spread quickly.

Researchers studying nanoparticles—including nanoplastics—have observed that under certain conditions, these particles can:

- interfere with mitochondrial function
- reduce energy production
- trigger stress responses inside the cell

This doesn't mean every exposure leads to damage.

But it does show that:

> **Mitochondria are a key point of vulnerability inside the cell.**

When they are disrupted, two major problems can occur.

1. Less Energy (The Slowdown Effect)

If mitochondria are not working properly, the cell produces less energy.

That means:

- slower repair
- reduced function
- decreased resilience

In simple terms:

> **The cell begins to run low on power.**

Over time, this can affect tissues that require high energy, such as:

- the brain
- the heart
- muscles

These are the systems that rely most heavily on efficient mitochondrial function.

2. More Oxidative Stress (The Damage Effect)

At the same time that energy production drops, another process can increase:

oxidative stress.

This is one of the most important—and widely studied—mechanisms in cellular damage.

To understand it, think of what happens when metal rusts.

Rust is the result of oxidation—a chemical reaction that gradually damages the material.

Something similar can happen inside cells.

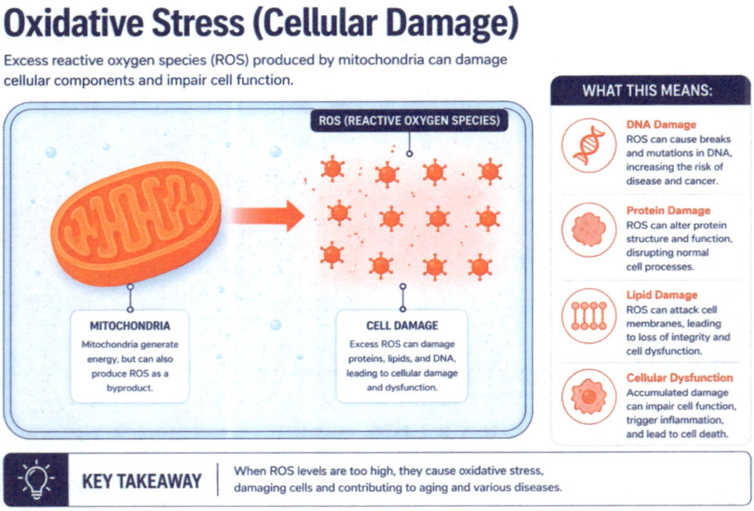

Cells naturally produce reactive molecules as part of normal metabolism. These are called **reactive oxygen species (ROS)**.

In small amounts, ROS are not only normal—they are necessary.

But when levels rise too high, they can begin to damage:

- cell membranes
- proteins
- DNA

This is what scientists refer to as **oxidative stress**.

A simple way to think about it:

Oxidative stress is like "cellular rusting."

Why Mitochondria Are Central to This Process

Mitochondria are both:

- a major source of energy
- and a major source of reactive oxygen species

When functioning normally, they maintain a balance.

But when disrupted:

- energy production drops
- ROS levels can increase

This creates a double effect:

- **less energy available to repair damage**
- **more damage being produced at the same time**

The system begins to work against itself.

Where Nanoplastics May Fit Into This

Research into nanoplastics is still developing, but studies on nanoparticles more broadly—and early studies on nanoplastics—suggest that:

- particles inside cells can contribute to increased oxidative stress
- they may interfere with mitochondrial function
- they may trigger cellular stress responses

This does not mean nanoplastics directly "attack" mitochondria in a simple or uniform way.

What it suggests is:

> **Their presence inside the cell may disrupt the balance that mitochondria depend on.**

And once that balance is disturbed, the effects can cascade.

A Compounding Problem

The concern is not just a single event.

It is the combination of factors:

- continuous exposure to nanoplastics
- the ability of particles to enter cells
- the sensitivity of mitochondria to disruption

Over time, even small effects—if repeated—may begin to add up.

This is one reason why researchers are increasingly focused on:

- long-term, low-level exposure

- cumulative effects rather than immediate toxicity

What We Know—and What We're Still Learning

What we know:

- Mitochondria are essential for energy production
- Disruption of mitochondrial function affects cell health
- Oxidative stress is a well-established mechanism of cellular damage
- Nanoparticles, including nanoplastics in some studies, can influence these processes

What we don't fully know:

- how often nanoplastics reach mitochondria in real human conditions
- how significant their impact is over long periods
- how different types of nanoplastics behave in different tissues

The Takeaway

Mitochondria are at the center of cellular life.

They provide the energy that keeps cells functioning—and help maintain the balance that prevents damage.

When that system is disrupted:

- energy decreases
- damage increases

Nanoplastics, once inside the cell, may contribute to that disruption—not as a single cause, but as part of a broader set of interacting factors.

And that is what makes this issue important.

> **At the cellular level, even small disruptions to energy and balance can have larger consequences over time.**

Understanding how and when that happens is one of the most important areas of ongoing research.

CHAPTER 6

Inflammation: The Body's Misfiring Defense

The human body is built to protect itself.

Every day, your immune system scans for threats:

- bacteria
- viruses
- damaged cells

When it detects something unfamiliar, it responds.

That response is called **inflammation**.

In the right situation, inflammation is not a problem.

It's a solution.

What Inflammation Is (And Why We Need It)

If you cut your skin, the area becomes:

- red
- swollen
- warm
- sometimes painful

That's inflammation at work.

It's the body:

- sending immune cells to the area

- increasing blood flow
- beginning the repair process

In short:

> **Inflammation is how the body defends and heals itself.**

Without it, even minor injuries or infections could become serious.

But like many systems in the body, it depends on **balance**.

When the System Stays Switched On

Inflammation is meant to be temporary.

It turns on when needed—and then turns off once the problem is resolved.

But sometimes, that "off switch" doesn't work properly.

When inflammation continues over time, it becomes:

chronic inflammation.

Instead of helping, it begins to cause damage.

This can affect:

- tissues
- blood vessels
- organs

And it has been linked to a wide range of long-term conditions.

How the Immune System Sees Nanoplastics

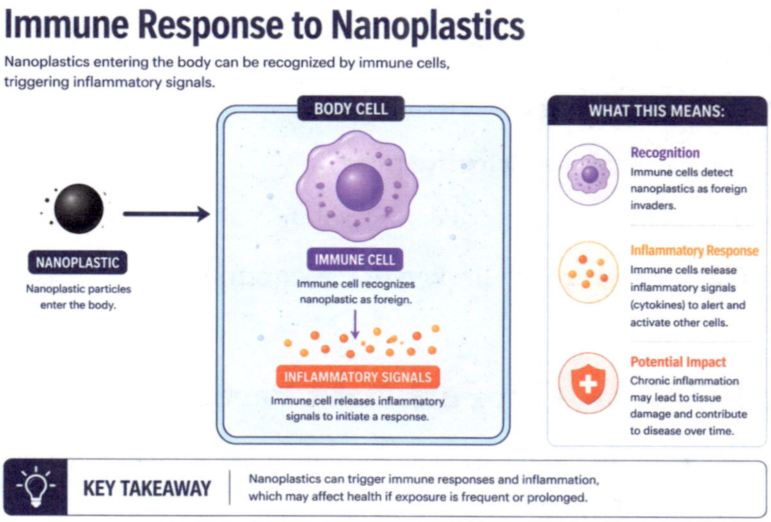

The immune system is not designed to recognize "plastic" as a category.

It responds to:

- size
- shape
- surface properties
- and whether something appears foreign or abnormal

Nanoplastics, once inside the body, can trigger attention because:

- they are foreign
- they may bind to proteins
- they may interact with cells in unusual ways

In response, immune cells may:

- attempt to isolate them
- try to break them down
- release signaling molecules

This is part of a normal defensive process.

But with nanoplastics, there's a complication:

They are not easily broken down or removed.

The Problem With Persistent Particles

When the immune system encounters something it cannot eliminate, it may continue to respond.

This can lead to:

- repeated activation of immune pathways
- ongoing signaling between cells
- sustained low-level inflammation

Unlike a clear infection that resolves, this becomes a **background condition**.

Not intense—but constant.

Chronic Inflammation (The Quiet Damage)

Chronic inflammation does not always feel dramatic.

There may be no obvious symptoms at first.

But over time, it can:

- interfere with normal cell function
- contribute to tissue damage
- increase stress on biological systems

Combined with other factors—like oxidative stress and mitochondrial disruption—it can become part of a larger pattern.

> **The body is no longer responding to a short-term threat. It is reacting continuously.**

Why Long-Term, Low-Level Exposure Matters

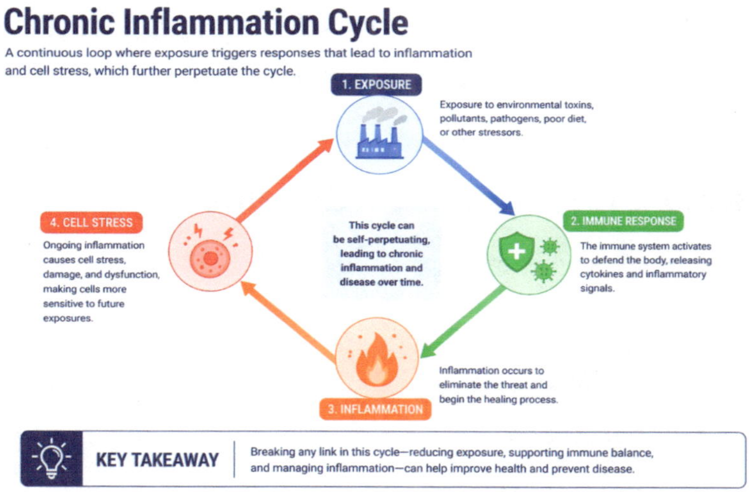

One of the biggest misconceptions about health risks is the idea that harm only comes from large exposures.

In reality, many biological systems respond to **repetition over time**.

With nanoplastics:

- exposure is frequent
- particles may enter the body repeatedly
- clearance may be incomplete

This creates a different kind of risk:

Not a single event—but a continuous input.

Each individual exposure may be small.

But together, over months and years, they may:

- maintain immune activation
- contribute to ongoing inflammation
- prevent the system from fully returning to baseline

A System Under Constant Pressure

When multiple processes overlap:

- cellular stress
- mitochondrial disruption
- immune activation

the effects can reinforce each other.

For example:

- oxidative stress can trigger inflammation
- inflammation can increase oxidative stress
- reduced cellular energy can make recovery more difficult

This creates a cycle.

Not necessarily immediate or visible—but persistent.

What We Know—and What We're Still Learning

What we know:

- Inflammation is a normal and necessary immune response
- Chronic inflammation can contribute to long-term health problems
- Nanoparticles, including nanoplastics in some studies, can trigger immune responses
- Persistent foreign materials can lead to sustained low-level inflammation

What we don't fully know:

- how strongly nanoplastics drive inflammation in real-world human exposure
- how long immune activation persists after exposure
- how different individuals respond over time

The Takeaway

Inflammation is not the enemy.

It is one of the body's most important defense systems.

But when that system is activated too often—or for too long—it can begin to work against the body it is meant to protect.

Nanoplastics introduce a new kind of challenge:

- they are small enough to enter the body
- foreign enough to trigger a response
- and persistent enough to keep that response active

Not as an immediate crisis—

but as a **continuous signal that never fully goes away.**

And over time, that signal may matter more than any single exposure ever could.

CHAPTER 7

The Brain Under Attack (Careful but Compelling)

The brain is not like other organs.

It controls movement, memory, emotion, and every system that keeps you alive. Because of this, the body protects it with one of the most selective defenses it has:

the **blood–brain barrier**.

This barrier exists for a simple reason:

> The brain is not supposed to be exposed to whatever is circulating in the bloodstream.

The Blood–Brain Barrier (The Brain's Security System)

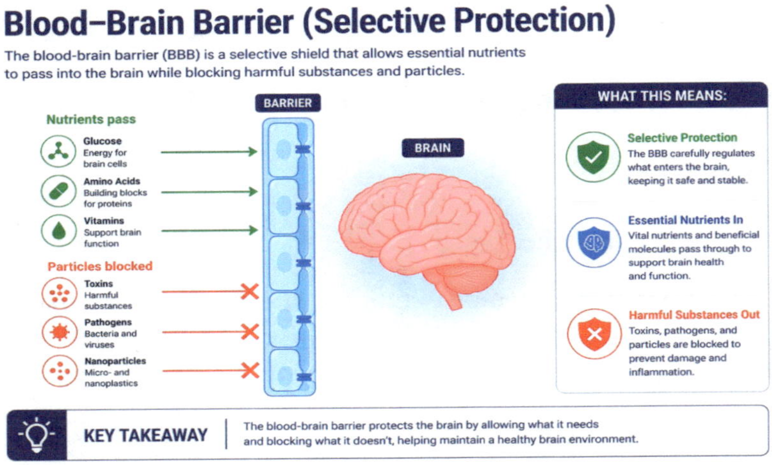

The blood–brain barrier (BBB) is a tightly controlled gateway between the blood and the brain.

It allows in what the brain needs:

- oxygen
- glucose
- essential nutrients

And it blocks most things it doesn't:

- toxins
- pathogens
- foreign particles

It is not just a physical wall. It is an active system that regulates what enters and what stays out.

In simple terms:

The brain is designed to be protected from outside contaminants.

Plastic in the Brain

Recent research examining human tissue has detected plastic particles in the brain at measurable levels.*

In these studies, brain tissue showed higher concentrations of plastic compared to other organs such as the liver and kidneys.

This is a critical point:

Plastic is not just entering the body—it is being detected in one of its most protected environments.

These findings shift the conversation.

The question is no longer whether exposure is happening.

It is how these particles are reaching the brain—and what happens once they are there.

How Could Particles Reach the Brain?

The blood–brain barrier is highly effective, but it is not absolute.

At the nanoscale, particles behave differently.

Because of their extremely small size, nanoplastics may:

- interact more closely with biological membranes
- be transported by cells within the body

- move through pathways that larger particles cannot use

This does not mean every particle crosses the barrier.

But it does mean:

> **The protective systems of the body may be challenged in ways they were never designed for.**

Why Size Matters

Larger particles are easier for the body to block or remove.

Nanoplastics are different.

They are small enough to:

- interact directly with cells
- bind to proteins and biological molecules
- potentially move across tightly controlled barriers

This is what makes them biologically relevant.

Not because they are large or obvious—

but because they are small enough to go where larger materials cannot.

The Brain's Communication System

The brain functions through a vast network of cells called neurons.

These cells communicate through connections known as **synapses**.

Every thought, memory, and action depends on these signals moving quickly and accurately.

This system depends on:

- stable cellular conditions
- balanced chemical signals
- efficient energy production

When that balance is disrupted, communication can become less effective.

Emerging Evidence of Disruption

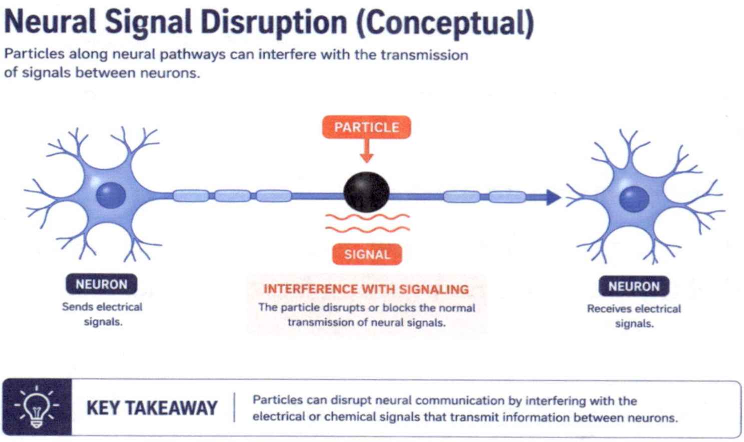

Research into nanoparticles has shown that, under certain conditions, these particles can:

- interfere with cellular signaling
- contribute to oxidative stress
- affect energy production within cells

These are the same processes already discussed:

- mitochondrial disruption
- oxidative stress
- inflammation

In the brain, these effects may be more significant because of how sensitive neural systems are.

A simplified way to understand this is:

> **If the signals between brain cells are disrupted, even slightly, the effects may accumulate over time.**

Cognition and Brain Function

Because the brain depends on precise communication, any disruption to that system raises important questions.

Researchers are investigating whether long-term exposure to nanoscale particles may influence:

- attention
- memory
- processing speed

At this stage, these are areas of ongoing study—not confirmed outcomes.

But they are being studied for a reason:

> **The brain is highly sensitive to small, persistent changes in its environment.**

Neurodegeneration — A Careful Discussion

Neurodegenerative conditions develop over long periods and involve many factors:

- aging
- genetics
- environmental influences
- cellular stress

Processes such as:

- chronic inflammation
- oxidative stress
- mitochondrial dysfunction

are already known to play a role in these conditions.

Because nanoplastics may contribute to these same processes, researchers are asking whether long-term exposure could be one contributing factor.

At this stage, there is no definitive answer.

But the question is no longer theoretical.

What We Know—and What We're Still Learning

What we know:

- The brain is protected by the blood–brain barrier
- Plastic particles have been detected in brain tissue*
- Nanoparticles can interact with biological barriers
- Cellular stress, inflammation, and oxidative damage affect brain function

What we don't fully know:

- how frequently nanoplastics reach the brain
- how long they remain once present
- how different types of particles behave in neural tissue
- the long-term effects of repeated exposure

The Takeaway

The brain is designed to be protected.

But nanoplastics introduce a new kind of challenge—particles small enough to interact with systems that were never meant to encounter them.

The presence of plastic in brain tissue changes the conversation.

> **This is not a question of whether exposure is happening.**
> **It is a question of what that exposure means over time.**

And when it comes to the brain, even small, persistent disruptions deserve serious attention.

Footnote

* Campen, M. J. et al. (2024). Analysis of human cadaver tissue (brain, liver, kidney) reported significantly higher concentrations of micro- and nanoplastic particles in brain samples compared to other organs, with measured levels in some cases approaching approximately 0.5% by weight. The study also identified polyethylene as the most common polymer detected, the most widely produced plastic in the world (used in bags, bottles, and packaging).

CHAPTER 8

Hormones, Organs, and System-Wide Effects

Up to this point, we've focused on what happens at the cellular level.

But the human body does not operate in isolation.

Cells form tissues.
Tissues form organs.
And organs work together as interconnected systems.

When something disrupts cells, the effects don't stay contained.

They spread.

The Body as a Connected System

Your body is constantly coordinating activity across multiple systems:

- the brain communicates with organs
- hormones regulate timing and balance
- the immune system monitors for threats
- the liver and kidneys process and remove substances

This coordination depends on stability.

When foreign materials enter the body—especially repeatedly—they can interact with more than one system at once.

Nanoplastics are not confined to a single pathway.

They move.

And wherever they move, they interact.

Hormones: The Body's Chemical Messengers

Hormones are the body's signaling system.

They control:

- growth
- metabolism
- sleep cycles
- reproduction
- stress response

They operate at extremely low concentrations, meaning:

Even small disruptions can have noticeable effects.

Some plastics and plastic-associated chemicals are known to interact with hormonal systems.

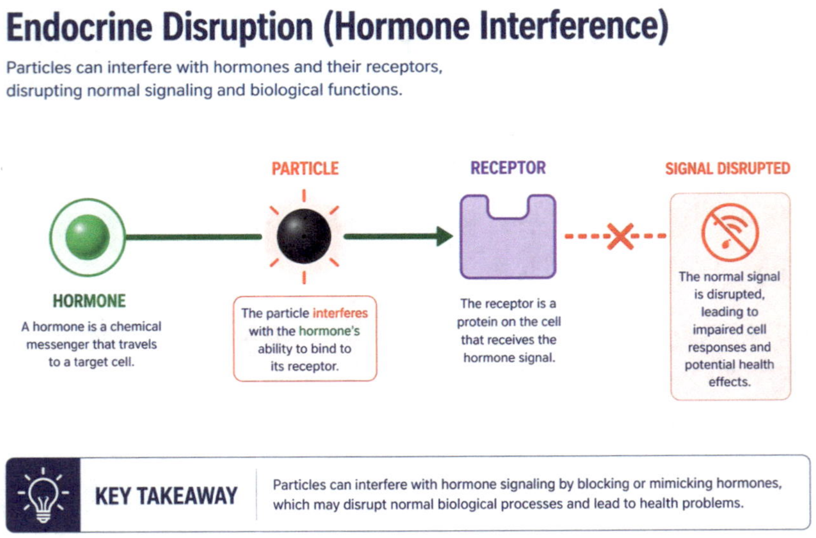

These are often referred to as **endocrine-disrupting chemicals (EDCs)**.

They may:

- mimic natural hormones
- block normal hormone signals
- alter how hormones are produced or regulated

In simple terms:

They can confuse the body's messaging system.

Nanoplastics and Hormonal Interference

Nanoplastics themselves are physical particles, but they are rarely "just plastic."

They can:

- carry chemical additives used during manufacturing
- absorb environmental pollutants
- bind to biological molecules once inside the body

This creates a complex interaction:

- the particle itself
- the chemicals attached to it
- and the biological environment it enters

Research is ongoing, but there is increasing interest in how nanoscale particles may:

- transport hormone-disrupting substances

- interact with endocrine tissues
- contribute to subtle imbalances over time

This is not always immediate or obvious.

Hormonal disruption often works slowly—through small changes that accumulate.

The Liver: The Body's Filter

The liver plays a central role in processing what enters the body.

It:

- filters blood
- breaks down chemicals
- prepares substances for removal

Because of this, it is one of the first organs exposed to foreign materials that enter through food and water.

If nanoplastics or associated compounds reach the liver, they may:

- be processed or modified
- interact with liver cells
- contribute to oxidative stress or inflammation

The liver is resilient.

But it is also continuously working.

Long-term exposure means long-term workload.

The Kidneys: The Body's Exit System

The kidneys filter waste from the bloodstream and regulate fluid balance.

They are designed to remove substances from the body efficiently.

However, removal depends on:

- particle size
- chemical properties
- how substances are transported in the blood

Some particles may be filtered and excreted.

Others may:

- persist in circulation
- accumulate in tissues
- or be only partially removed

This creates uncertainty around how effectively the body can eliminate nanoplastics over time.

The Gut: The Front Line of Exposure

The digestive system is one of the main entry points for environmental exposure.

Every day, it encounters:

- food
- water
- packaging-related residues

- environmental contaminants

The gut is not just a passive tube.

It is an active system that:

- absorbs nutrients
- interacts with the immune system
- hosts a complex microbiome

Nanoplastics introduced through ingestion may:

- interact with the gut lining
- influence immune responses
- affect the balance of gut bacteria

This is an emerging area of research, but it is important because:

> **The gut connects directly to both the immune system and overall health.**

Accumulation: A Growing Concern

One of the key questions surrounding nanoplastics is whether they accumulate in the body over time.

We already know:

- exposure is ongoing
- particles can enter the body
- some are detected in tissues

What remains uncertain is:

- how quickly they build up

- how evenly they are distributed
- how effectively they are cleared

But the concern comes from a simple pattern:

> **If input continues and removal is incomplete, accumulation becomes possible.**

Even slow accumulation, over years or decades, could change the internal environment of the body.

System-Wide Effects

When multiple systems are affected at once, the impact is not isolated.

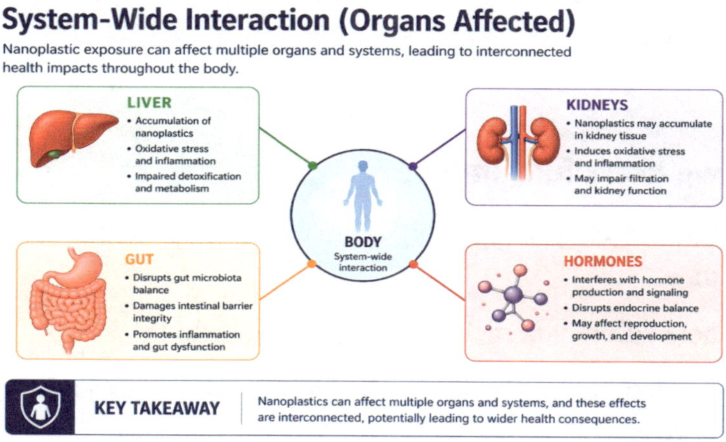

Instead, effects can overlap:

- hormonal imbalance can affect metabolism and mood
- liver stress can influence detoxification capacity
- kidney function affects waste removal

- gut disruption can influence immune signaling

These systems do not operate independently.

They reinforce—or strain—each other.

What We Know—and What We're Still Learning

What we know:

- Some plastics and their additives can disrupt hormonal systems
- The liver, kidneys, and gut are primary points of interaction
- Nanoparticles can move through the body and interact with multiple tissues
- Repeated exposure increases the importance of long-term effects

What we don't fully know:

- how much nanoplastics accumulate in different organs over time
- how different particle types behave biologically
- the long-term impact of low-level, chronic exposure across systems

The Takeaway

The body is not a collection of isolated parts.

It is a network.

When foreign materials enter that network—especially continuously—they have the potential to influence multiple systems at once.

Nanoplastics are not just a localized issue.

They are a **system-wide variable**.

> **And when multiple systems are affected, even small disruptions can add up to something larger over time.**

Part 3
Accumulation & Long-Term Risk

CHAPTER 9

Do Nanoplastics Build Up in the Body?

If exposure were rare, this would be a smaller problem.

But exposure is not rare.

It is daily.

So the question that matters most isn't just *whether nanoplastics can get in.*

It's whether the body can **get them back out**—fast enough to prevent buildup.

The Core Question: Input vs Removal

Your body is constantly managing foreign material.

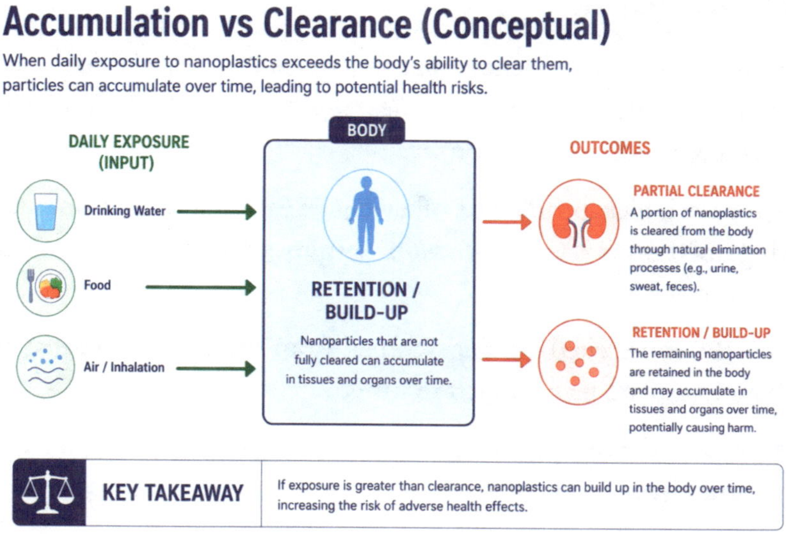

Some things are:

- filtered by the kidneys
- processed by the liver
- eliminated through bile, urine, and stool

But nanoplastics introduce a unique challenge:

- they can be extremely small
- they can bind to proteins and "blend in" biologically
- they are not designed to break down like natural materials

So the bioaccumulation question comes down to a simple imbalance:

> **If exposure continues and clearance is incomplete, buildup becomes possible.**

What We Know

We now have real evidence that micro- and nanoplastics can be detected in multiple human tissues and organs (including blood, placenta, liver, kidney, and brain).[2]

We also have a high-profile line of evidence from human autopsy tissue showing patterns consistent with **organ-specific accumulation**, especially in the brain.

One of the most cited findings comes from research associated with Matthew Campen's team, which reported:

- **brain tissue contained far higher plastic concentrations than liver or kidney**

- the detected material included particles often described as **micro- and nanoplastics**
- **polyethylene** was the dominant polymer
- and measured concentrations were higher in the more recent (2024) samples compared with earlier sampling (2016), consistent with an upward trend over time.[1]

Those results have driven serious attention because they point to something more than "temporary exposure":

> **They suggest retention and selective buildup in certain tissues.**[1]

What We Don't Know (Yet)

Even with strong studies, several critical things remain uncertain:

- **How fast** accumulation occurs for different people
- **How long** particles persist once embedded in tissue
- **What fraction** are cleared versus retained
- **Which particle types** (size, shape, polymer, additives) are most persistent
- **How much** comes from lifetime exposure versus recent exposure

And there's an important scientific reality you should acknowledge in this book:

Detection methods for micro- and nanoplastics in human tissue are improving rapidly, but the field still faces challenges around contamination control, standardization, and interpreting results—especially in complex tissues.[3,4]

That does **not** mean "it's fake."

It means this is an emerging science moving from "we found it" to "we can measure it reliably everywhere, the same way, every time."

Bioaccumulation Evidence: Early but Concerning

Beyond the Campen-associated work, other post-mortem research continues to report **heterogeneous, organ-specific accumulation** patterns—meaning some tissues show higher loads than others, consistent with selective retention rather than uniform distribution.[5]

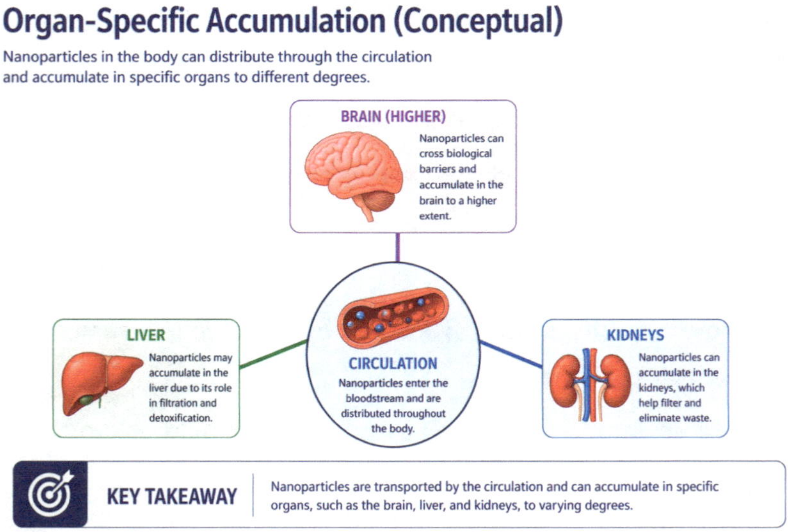

This is exactly what you would expect if:

- certain organs act as long-term sinks
- certain particle sizes or shapes are harder to clear
- and exposure continues for years

In other words, the concern isn't just that particles can enter.

It's that the body may not be able to fully reverse the process once they do.

Why Chronic Exposure Matters More Than Single Exposure

A single exposure event matters less than a lifetime pattern.

Many health risks don't come from one large dose.

They come from small inputs that never stop.

Nanoplastics fit that profile because:

- exposure is continuous (air, water, food, indoor dust, synthetic materials)
- entry pathways are multiple
- and clearance may be incomplete or variable

This is why chronic exposure changes the risk conversation:

> **Even if the body clears most particles, "most" is not the same as "all."**
> **If a small fraction remains each day, buildup becomes possible over time.**

And because organ systems are connected:

- retained particles can contribute to inflammation
- inflammation can increase biological stress
- stress can make tissues more vulnerable
- and vulnerable tissues may handle future exposures less effectively

This is how long-term risk grows—not through a single dramatic event, but through cumulative pressure.

What This Means for You

The most honest message is also the most compelling:

- **We already have evidence of plastic particles in human tissues.**[2]
- **We have evidence consistent with selective accumulation, especially in the brain, and signals that concentrations may be rising over time.**[1]
- **What we do not yet have is a full "personal lifetime map" showing exactly how much any one person accumulates and what it means for their long-term health.**[3]

So the uncertainty is not whether this is real.

The uncertainty is the *individual trajectory*:

> How much builds up in *your* body, in *your* organs, across *your* lifetime.

That is why this chapter matters.

Because it reframes the issue:

Not as science fiction.

As an exposure and retention problem that modern research is now measuring directly—while still working out the long-term consequences.

Footnotes

[1] **Nihart AJ, et al.** "Bioaccumulation of microplastics in decedent human brains." *Nature Medicine* (2024/2025 online record). (Nature)

[2] **Roslan NS, et al.** "Detection of microplastics in human tissues and organs" (scoping review, 2024). (PMC)

[3] **Science Media Centre** expert reactions highlighting methodological questions and the need for clearer explanations/controls in organ accumulation studies (2025). (Science Media Centre)

[4] Ongoing public scientific debate over measurement reliability/false positives in parts of the tissue literature (context on why careful framing builds credibility). (The Guardian)

[5] **Dzierżyński E, et al.** "Post-mortem evidence of microplastic bioaccumulation in ..." (2025 autopsy evidence of organ-specific accumulation). (PMC)

CHAPTER 10

The Generational Risk

Up to this point, the focus has been on what nanoplastics may do to the human body over time.

But there is another layer to this issue—one that extends beyond individual exposure.

It is the possibility that exposure does not start in adulthood.

It may begin **before birth**.

The Placenta: A Critical Barrier

During pregnancy, the placenta serves as the interface between mother and developing child.

It performs essential functions:

- delivering oxygen and nutrients
- removing waste
- regulating what passes between maternal and fetal blood

Like the blood–brain barrier, it is selective.

It is designed to protect the developing fetus from harmful substances.

But it is not an impenetrable wall.

Evidence of Transfer

The idea that plastic particles can reach the developing fetus is no longer theoretical.

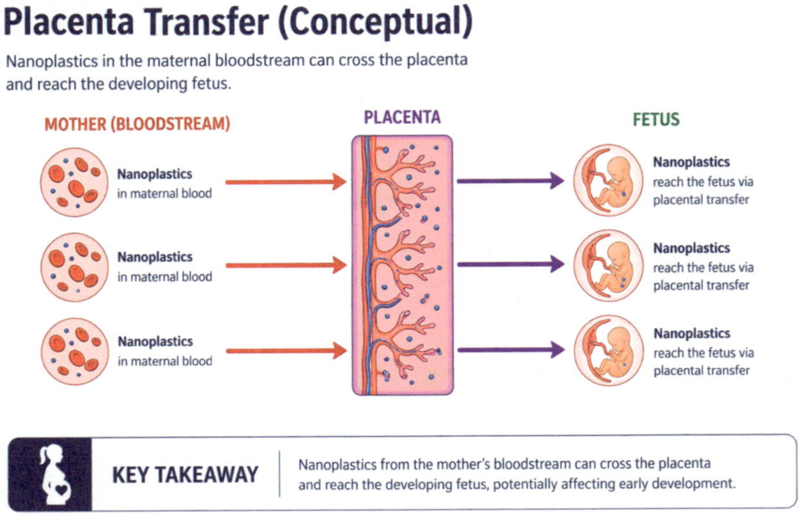

A landmark study conducted in Italy examined human placentas using advanced analytical techniques.

What researchers found was significant:

- **Microplastic particles were detected in human placental tissue**[1]
- The particles were found in multiple regions, including:
    - the **maternal side**
    - the **fetal side**
    - and the **membranes surrounding the fetus**[1]

This means the particles were not confined to one area—they were distributed across the entire interface between mother and child.

In simple terms:

> **Plastic particles were present in the very structure designed to protect and support fetal development.**

More recent research has extended these findings.

Studies have now reported microplastic particles in:

- **placenta tissue**
- **umbilical cord blood**
- **and **meconium (a newborn’s first stool)**[2]**

These findings suggest:

> **Exposure may begin before birth—and may reach the developing fetus itself.**

Why Early-Life Exposure Matters

Early development is one of the most sensitive periods in human life.

During this time:

- organs are forming
- the brain is developing rapidly
- cellular systems are establishing long-term patterns

Small disruptions during this stage can have effects that extend far beyond infancy.

This is a well-established principle in biology:

> **The earlier the exposure, the greater the potential impact.**

A Developing System, Not a Finished One

Adults have fully formed systems.

Children do not.

In early life:

- the immune system is still developing
- detoxification systems are not fully mature
- protective barriers may function differently than in adults

This means the same exposure can have different consequences depending on when it occurs.

A level that an adult body can manage may be more significant for a developing system.

Why Children May Be More Vulnerable

There are several reasons why children may be more affected by environmental exposures:

- **Higher relative intake**
  Children breathe more air, drink more water, and eat more food per unit of body weight than adults.

- **Developing organs**
  Systems such as the brain, endocrine system, and immune system are still forming.
- **Longer lifetime exposure**
  Early exposure extends the total duration over which accumulation and effects can occur.
- **Behavioral factors**
  Hand-to-mouth activity and close contact with surfaces can increase exposure to environmental particles.

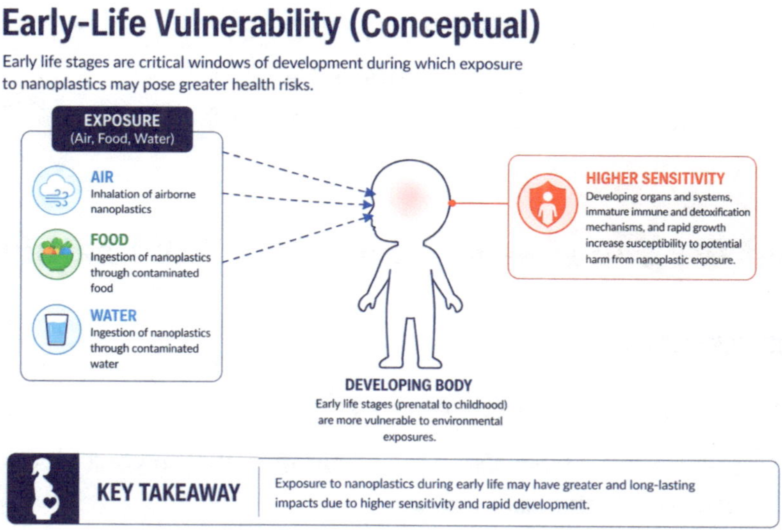

Together, these factors create a different risk profile.

Not necessarily immediate—

but potentially more significant over time.

The Compounding Effect

When exposure begins early, it doesn't just add to lifetime exposure—it extends it.

This means:

- more time for accumulation
- more time for low-level biological stress
- more time for small effects to build

And because development is ongoing, early exposure may influence how systems function later in life.

> **This is not just exposure during life.**
> **It is exposure at the very beginning of life.**

What We Know—and What We're Still Learning

What we know:

- Microplastic particles have been detected in placental tissue[1]
- Microplastics have also been found in umbilical cord blood and newborn biological samples[2]
- The placenta allows selective transfer between mother and fetus
- Early development is a sensitive period for environmental exposure
- Children have different exposure patterns and vulnerabilities than adults

What we don't fully know:

- how frequently nanoplastics cross into the fetal environment
- how much accumulates during development
- the long-term effects of prenatal and early-life exposure
- how these exposures interact with other environmental and genetic factors

The Takeaway

The concern with nanoplastics is not limited to what they may do to adults over time.

It is also about when exposure begins.

If exposure starts before birth and continues throughout life, the timeline of interaction with the body becomes much longer.

> **This is not just an individual issue.**
> **It is a generational one.**

Not because we have all the answers—

but because the combination of:

- early exposure
- continuous exposure
- and uncertain long-term effects

creates a question that cannot be ignored.

And the earlier that timeline begins, the more important that question becomes.

Footnotes

[1] Ragusa, A. et al. (2021). *Plasticenta: First evidence of microplastics in human placenta.* Microplastics were detected in multiple regions of the placenta, including maternal and fetal sides and membranes.

[2] Subsequent studies (2023–2024) have reported microplastics in placenta, umbilical cord blood, and meconium, indicating transfer along the maternal–fetal pathway.

PART 4
The Detox Question

CHAPTER 11

Can You Detox Nanoplastics? (Truth, Not Hype)

This is where scams thrive.

When you feel threatened, the natural response is to look for a solution—fast. And the internet is full of expensive programs claiming to "pull plastics out of your body."

Here's the honest reality:

There is currently no clinically proven "nanoplastic detox" that reliably removes micro- or nanoplastics from human tissues on demand.[1,2]

Anyone selling certainty here is selling marketing, not medicine.[3]

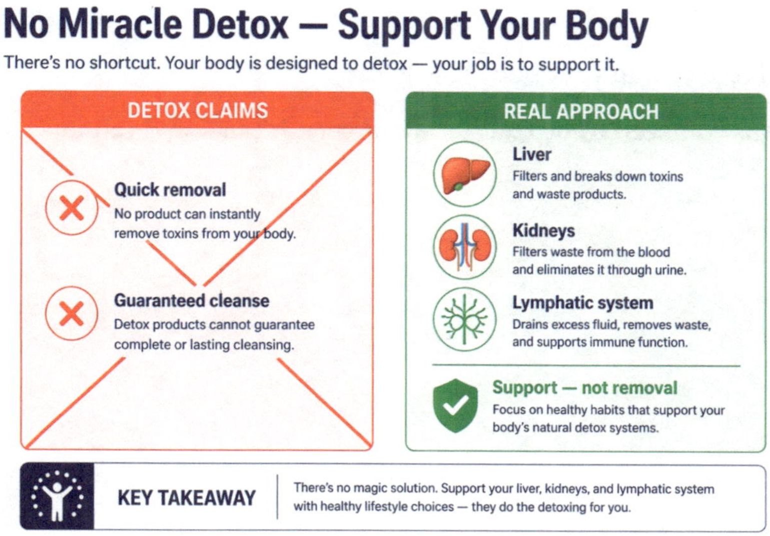

What "Detox" Actually Means (In the Body)

Your body already has systems designed to manage foreign substances:

The liver
Processes and modifies many chemicals so they can be eliminated.

The kidneys
Filter blood and remove waste through urine.

The lymphatic system
Moves fluids, immune cells, and cellular waste products. It supports immune surveillance and cleanup.

These systems are real. They are working all the time.

But there is a key limitation:

Your detox systems evolved to handle biological substances and many chemicals.
They did not evolve to solve modern synthetic particles at the nanoscale.

That doesn't mean your body is helpless.

It means you need to be realistic about what it can—and cannot—do.

What We Know About Elimination (What Leaves the Body)

For plastics that enter through food and drink, one clear route out is simple:

feces.

Microplastics have been detected in human stool repeatedly, confirming that at least some ingested particles pass through and are eliminated.[4]

A 2024 human intervention pilot study tested different diet-related exposure scenarios and measured microplastics in stool, supporting the idea that stool reflects both exposure and elimination.[5]

There is also evidence that plastics—or particles within certain size ranges—can appear in urine, based on human biomonitoring studies using spectroscopic methods.[6,7]

So the honest conclusion is:

Some plastics clearly leave the body through stool, and there is evidence suggesting elimination through urine for certain particle sizes or forms.[4,6]

What does not exist is a proven way to flush out plastics that have already embedded in tissues.

The "Support Your Body" Strategy (Not Miracle Removal)

If no magic detox exists, what can you actually do?

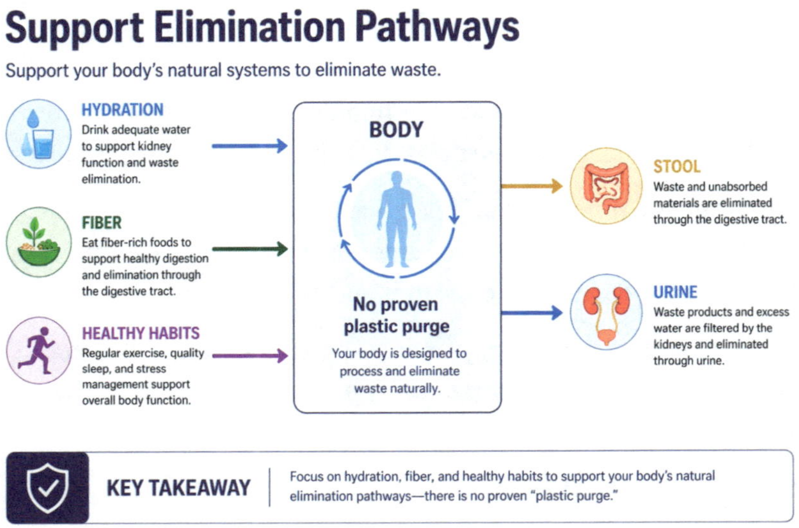

Focus on two things:

1. Reduce ongoing exposure (this becomes critical going forward)
2. Support your body's normal elimination pathways

This is about lowering your overall burden over time—not pretending you can erase it in a weekend.

Hydration (Helps the System Work, Doesn't "Wash Out Plastic")

Hydration supports kidney function and overall elimination.

But be precise:

Drinking more water does not remove plastic from your tissues. It supports normal physiology—which matters when exposure is ongoing.

Fiber (Your Most Credible Tool)

Fiber matters because it works where much of the exposure occurs: the gut.

Fiber can:

- increase stool bulk and transit
- bind and carry some substances out of the digestive tract
- support gut barrier function and microbiome health

There are no definitive clinical trials showing that fiber removes nanoplastics from organs.

But there is evidence pointing in a useful direction:

In animal research, certain indigestible fibers—especially chitosan—have been shown to increase fecal excretion of ingested polyethylene microplastics and reduce intestinal retention.[8]

Early human research has explored chitosan-based supplementation and measured changes in fecal microplastic findings after standardized exposure, though this remains preliminary and is not a proven therapy.[9]

So the correct way to understand this is:

Fiber is a reasonable, low-risk way to support gut elimination and gut health.
It may help reduce what gets absorbed and support what leaves through stool.
It is not a "plastic purge."[8,9]

Sweating (Be Careful Here)

This is where detox claims become the most exaggerated.

Here's what the science actually supports:

- Chemicals associated with plastics—such as certain additives—can be found in sweat or can interact with sweat at the skin surface.[10,11]
- There is no strong evidence that sweating removes intact microplastic or nanoplastic particles from inside the body

In fact, some research suggests sweat can leach additives out of microplastics, and those chemicals may then be absorbed through the skin—meaning sweating may sometimes increase chemical exposure under certain conditions.[10,12]

So the realistic takeaway is:

Sweating has benefits.
But it is not a method for removing plastic particles from the body.

Treat sauna and sweat claims as possible modifiers of chemical exposure—not as plastic removal strategies.

What This Means for You

You don't need to believe in miracles to take action.

Here's what actually holds up:

- There is no proven nanoplastic detox.[1,2,3]
- Your body does have real elimination systems, and some plastics do leave through stool, with evidence suggesting urine elimination for certain forms.[4,6]
- The most defensible strategy is long-term:
  - reduce exposure
  - support gut elimination (fiber)
  - support kidney and liver function (hydration, overall health)
  - avoid detox programs that make claims the science cannot support[3]

The goal is not fantasy removal.

It's reducing what goes in—and supporting what goes out.

> Support your body.
> Don't buy miracle removal.

Footnotes

[1] Ali, N. et al. (2025). Review of micro- and nanoplastics exposure and associated disease risks, highlighting key limitations and evidence gaps. *The Lancet Planetary Health.* (The Lancet)

[2] Sadique, S.A. et al. (2025). Review noting limited conclusive evidence for specific human outcomes and ongoing uncertainty about internal kinetics/clearance. *ScienceDirect (review).* (ScienceDirect)

[3] Reporting and expert warnings that commercial "microplastics detox" treatments lack evidence and may be irresponsible. (The Guardian)

[4] Hartmann, C. et al. (2024). Human stool intervention pilot measuring microplastics under different diet/plastic-use scenarios. *Science of the Total Environment.* (ScienceDirect)

[5] Austrian Environmental Agency summary referencing stool detection across participants and common polymers found. (umweltbundesamt.at)

[6] Pironti, C. et al. (2022). Preliminary report of microplastics detected in human urine using Raman microspectroscopy. *PMC.* (PMC)

[7] Ji, S. et al. (2025). Quantitative assessment of micro-/nanoplastics (≥300 nm) in non-invasive human samples including urine. *MDPI Toxics.* (MDPI)

[8] Liu, D. et al. (2025). Chitosan promoted fecal excretion of ingested polyethylene microplastics and reduced intestinal retention in rats. *Scientific Reports.* (Nature)

[9] Casella, C. et al. (2025). Preliminary human crossover study investigating chitosan supplementation and fecal microplastic findings after standardized meal exposure. *MDPI Foods.* (MDPI)

[10] University of Birmingham reporting on experimental evidence that sweat can leach additives from microplastics. (University of Birmingham)

[11] Genuis, S.J. et al. (2011). Human biomonitoring detecting BPA in sweat (chemical exposure marker; not plastic particles). *PMC.* (PMC)

[12] Abafe, O.A. et al. (2024). 3D human skin-equivalent model work on dermal bioavailability of additives present in microplastics. *Environment International.* (ScienceDirect)

PART 5

Protection & Risk Reduction

CHAPTER 12

Reducing Exposure at Home

You don't need a perfect solution to make a real difference.

You just need to reduce what comes in.

Nanoplastics are not a one-time exposure problem. They are a **daily input problem**—air, water, food, and the materials around you. That means the most effective strategy is simple:

> **Reduce exposure where you live.**

Your home is the one place you actually control.

Start With the Biggest Win: Don't Heat Plastic

Heat changes everything.

When plastic is exposed to high temperatures—microwaving, hot liquids, dishwashers—it can:

- release more particles
- break down faster
- shed both micro- and nanoplastics into food or liquids

This is one of the most avoidable sources of exposure.

What to do instead:

- Don't microwave food in plastic containers

- Avoid pouring hot liquids into plastic (including takeout containers)
- Replace plastic food storage with glass or stainless steel
- Be cautious with single-use packaging that contacts hot food

If it's hot, it shouldn't touch plastic.

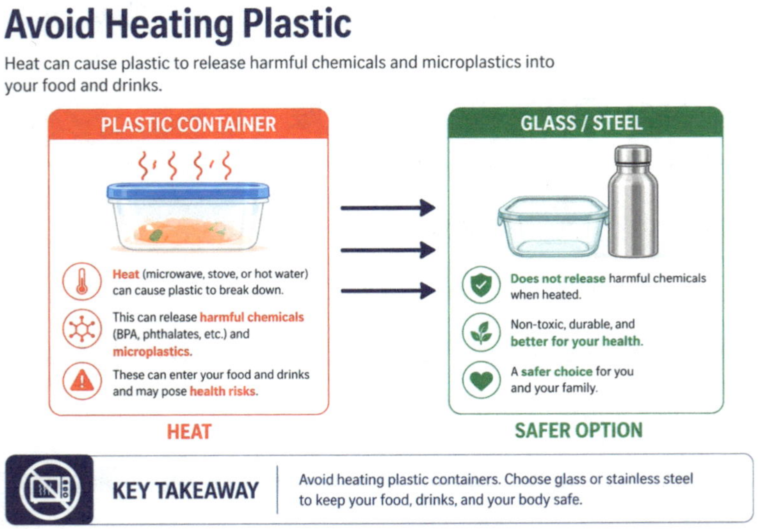

That one change alone can significantly reduce direct ingestion.

Water Matters More Than You Think

Water is a constant exposure pathway.

Even treated drinking water can contain microplastics—and potentially nanoplastics, which are harder to measure and remove.

The good news:

You can improve this.

Reverse Osmosis (RO)

- One of the most effective household filtration methods
- Can remove a large range of contaminants, including very small particles
- More expensive, but high performance

Activated Carbon Filters

- Common (pitcher filters, faucet filters)
- Effective for chemicals and some particulates
- Less reliable for the smallest particles compared to RO

No system is perfect.

But filtration reduces what you consume every day—and daily reductions add up.

> **You don't need zero exposure. You need less exposure, consistently.**

Your Clothes Are a Source You Don't See

Synthetic fabrics shed.

Every time you:

- wear them
- wash them

- dry them

they release tiny fibers into the air and your environment.

Over time, those fibers break down further—into smaller and smaller particles.

Indoors, this becomes part of what you breathe.

What helps:

- Increase use of natural fibers (cotton, wool, linen)
- Avoid over-reliance on synthetic clothing when possible
- Wash synthetic fabrics less aggressively (gentler cycles)
- Use lint filters or fiber-catching laundry solutions if available

You don't need to throw everything away.

But reducing synthetic load in your immediate environment reduces airborne exposure.

Dust: The Overlooked Source

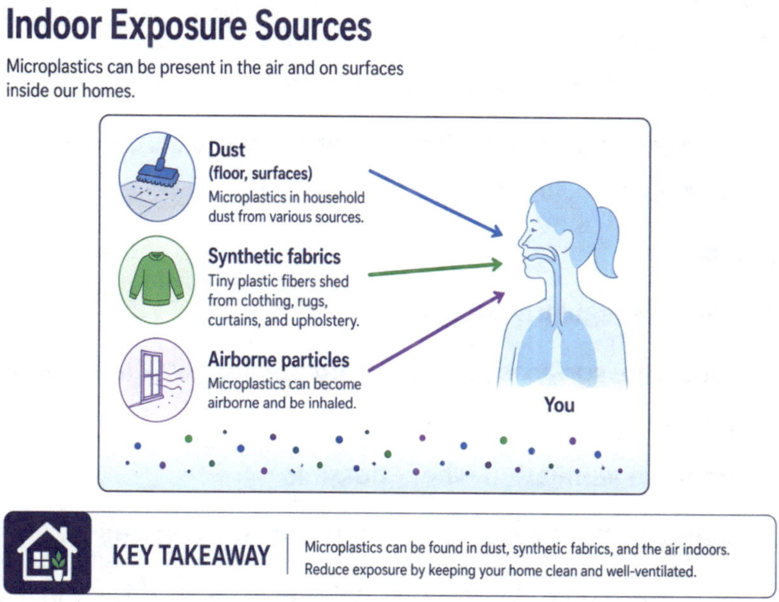

This is one of the most important—and least understood—sources of exposure.

Indoor dust is not just dirt.

It's a mixture of:

- fabric fibers
- degraded plastics
- particles from furniture, packaging, and household materials

And it doesn't stay on the floor.

It becomes airborne.

You breathe it.
It settles on food.
Children come into contact with it constantly.

> **For many people, dust is a primary exposure pathway —not an occasional one.**

What helps:

- Vacuum regularly using a HEPA filter
- Wet wipe surfaces instead of dry dusting (which redistributes particles)
- Improve ventilation where possible
- Remove shoes indoors to reduce outside contaminants
- Pay attention to bedrooms—where you spend the most time

Small habits here have a large impact over time.

You Don't Need Perfection—You Need Direction

Trying to eliminate all plastic exposure is unrealistic.

Trying to reduce it is not.

The goal is not zero.

The goal is **less, every day**.

- Less plastic in contact with heat
- Less in your water
- Less in your air
- Less in your immediate environment

Each change is small.

Together, they shift your long-term exposure in a meaningful way.

The Takeaway

You are not stuck with the environment as it is.

Inside your home, you have control.

And control—used consistently—is powerful.

Reduce what comes in.
Support what goes out.
Repeat it daily.

That's not a quick fix.

But it's what works.

CHAPTER 13

Food and Packaging Choices

Every time you eat, you are making more than a nutritional choice.

You are also making an exposure choice.

Food is one of the most direct ways nanoplastics can enter the body—not just because of what you eat, but because of **what your food touches before it reaches you**.

That means reducing exposure isn't just about ingredients.

It's about **contact.**

The Hidden Factor: Contact Matters

Plastic doesn't need to be visible to matter.

It can:

- shed particles
- transfer chemicals
- break down under heat, pressure, or time

And food is an ideal carrier.

It's moist.
It's often warm.
It sits in contact with packaging for extended periods.

If plastic touches your food, there is a pathway.

The goal is to reduce how often that happens—especially under conditions that accelerate release.

Hot Food + Plastic = Higher Exposure

This is one of the most consistent patterns across research and real-world observation:

Heat increases release.

When hot food comes into contact with plastic:

- the material softens
- bonds weaken
- particles and additives are more likely to transfer

This applies to:

- takeout containers
- microwave meals
- hot beverages in plastic-lined cups
- reheating food in plastic storage

Even “microwave-safe” plastic does not mean inert.

It means it won’t melt under certain conditions—not that it won’t release particles.

The hotter the contact, the higher the risk.

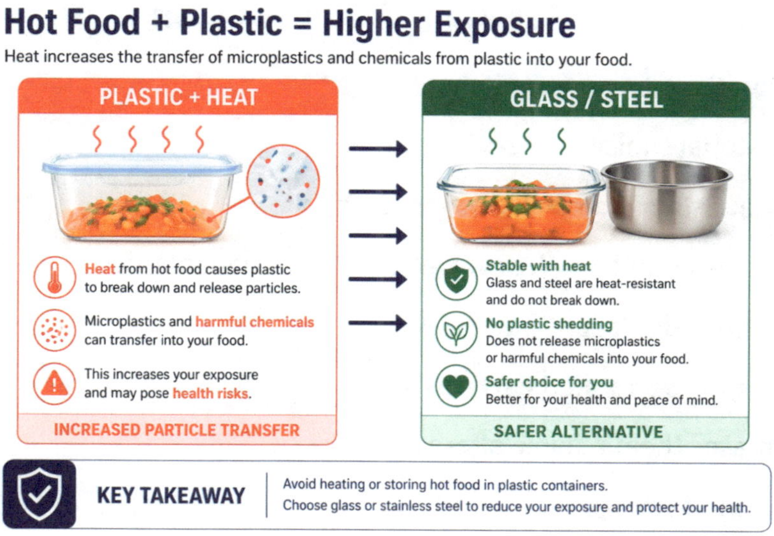

Simple changes matter:

- transfer hot food out of plastic before eating
- avoid microwaving in plastic containers
- be cautious with hot drinks in plastic or plastic-lined packaging

You don't have to eliminate every exposure.

But reducing heat-related contact is one of the easiest ways to lower intake.

Choose Better Materials When You Can

You don't need a perfect kitchen.

But material choice matters over time.

Glass and stainless steel are more stable:

- they don't break down into micro- or nanoplastics
- they don't soften with heat
- they don't shed particles into food

That makes them reliable alternatives for:

- food storage
- reheating
- drinking containers

Even partial changes help:

- switching your most-used containers
- replacing items that regularly contact hot food
- using glass for leftovers and reheating

You don't need to replace everything.
Just start with what you use the most.

Ultra-Processed Foods: The Overlooked Multiplier

Ultra-processed foods increase exposure in multiple ways.

They are:

- heavily packaged
- mechanically processed
- often heated during production
- stored for longer periods

That means more contact points:

- mixing equipment
- packaging lines
- storage materials
- transport environments

Each step adds potential for particle transfer.

This doesn't mean every packaged food is unsafe.

But it does mean:

> **The more processed and packaged the food, the more opportunities for exposure.**

A simple shift helps:

- increase whole or minimally processed foods
- reduce reliance on heavily packaged items
- pay attention to how food is stored and served

This aligns with general health guidance—and it also reduces plastic exposure.

Small Changes, Real Impact

You don't need extreme measures.

You need consistency.

- Move hot food out of plastic
- Use more stable materials when possible
- Reduce highly processed, heavily packaged foods

Each change reduces one pathway.

Together, they reduce your overall exposure.

If You Want to Go Further

If you're ready to take this beyond small changes, there is a deeper path.

Reducing exposure in food and packaging is just one part of a larger system—your home environment.

Everything from:

- kitchen materials
- storage habits
- cleaning products
- furniture and fabrics

plays a role.

That's exactly what is explored in more detail in *The Plastic Free Home*—a practical guide to reducing everyday exposure across the entire living environment.

This chapter gives you the foundation.

That book shows you how far you can take it.

The Takeaway

Food is not just what you eat.

It's what your food has touched.

Reduce contact.
Reduce heat exposure.
Choose better materials when you can.

You don't need perfection.

You need fewer pathways—and better habits over time.

CHAPTER 14

Air Matters More Than You Think

When people think about exposure, they usually think about food or water.

Air is often overlooked.

It shouldn't be.

You breathe every minute of your life.

That makes air one of the most consistent—and underestimated—exposure pathways for micro- and nanoplastics.

And most of that exposure doesn't happen outside.

It happens inside your home.

Indoor Air: The Invisible Source

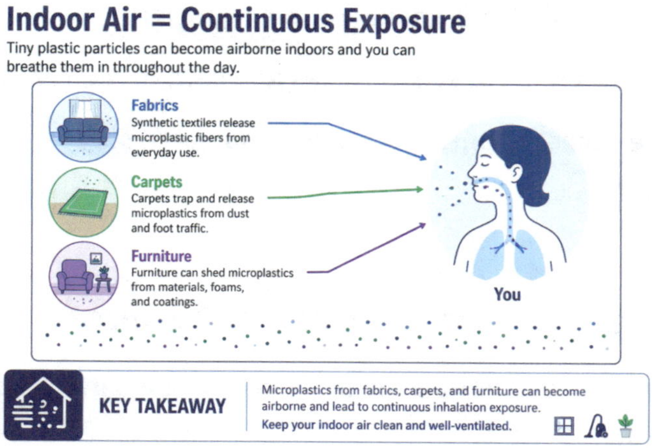

Indoor air is not empty.

It contains a mix of particles from:

- fabrics
- carpets
- furniture
- packaging materials
- everyday household items

Many of these are synthetic.

Over time, they break down and release tiny fibers and fragments into the air.

These particles:

- float
- settle
- get stirred back up
- and are inhaled without you ever noticing

You don't have to see particles for them to matter.

In fact, the smallest ones—the ones you *can't* see—are often the most relevant.

Microfiber Shedding: A Constant Source

One of the biggest contributors to indoor plastic particles is clothing and textiles.

Synthetic fabrics—like polyester, nylon, and acrylic—shed continuously.

Every time you:

- move
- sit
- walk across a carpet
- use bedding or upholstery

microscopic fibers are released into the environment.

Those fibers don't just stay on surfaces.

They become airborne.

And once airborne, they become inhalable.

Over time, larger fibers break down into smaller fragments—including particles in the nano range.

> **Your environment is constantly generating new particles.**

Not because something is wrong—but because of what it's made from.

Ventilation Changes the Equation

Air that doesn't move accumulates particles.

Air that moves dilutes them.

Ventilation is one of the simplest and most effective ways to reduce indoor particle concentration.

What helps:

- opening windows when conditions allow
- increasing fresh air exchange
- using exhaust fans in kitchens and bathrooms
- avoiding sealed, stagnant environments for long periods

Even small increases in airflow can reduce particle buildup.

Still air concentrates exposure.
Moving air reduces it.

HEPA Filtration: Removing What You Can't See

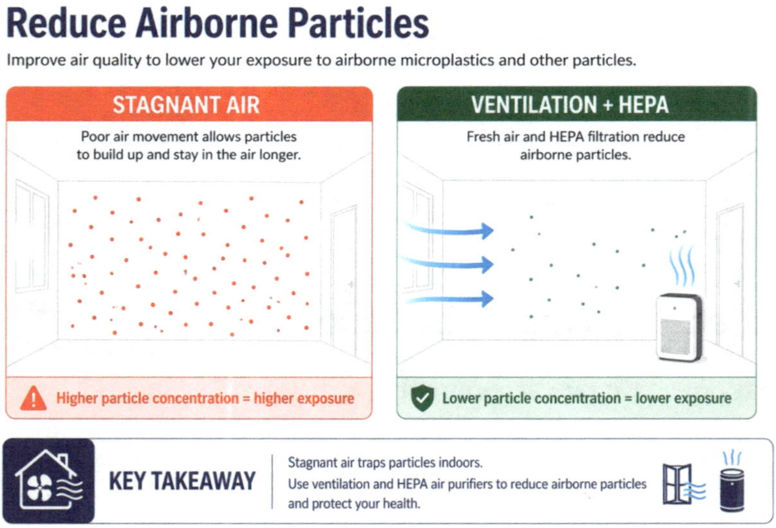

Not all particles settle.

Many remain suspended in the air—especially the smaller ones.

This is where filtration becomes important.

HEPA filters (High-Efficiency Particulate Air filters):

- are designed to capture very small particles
- are widely used in medical and clean-air environments
- can significantly reduce airborne particulate levels indoors

They don't eliminate everything.

But they reduce the overall load.

That matters—because exposure is cumulative.

Placement matters too:

- bedrooms (where you spend hours breathing continuously)
- main living areas
- enclosed spaces with limited airflow

You can't remove every particle.
But you can reduce what you breathe.

Reducing Microfiber Load Indoors

Filtration helps.

But reducing the source matters just as much.

Ways to lower microfiber shedding indoors:

- increase use of natural fibers (cotton, wool, linen)
- reduce reliance on synthetic textiles where practical
- wash synthetic clothing less aggressively
- avoid over-drying synthetic fabrics (which increases breakdown)
- clean surfaces regularly using methods that capture particles (not just move them)

Again, this is not about perfection.

It's about direction.

If You Want to Go Further

Air quality is only one part of a larger system.

The materials in your home—what you wear, sit on, sleep on, and store food in—all contribute to your overall exposure.

If you want a deeper, structured approach to reducing that exposure across your entire living environment, *The Plastic Free Home* explores:

- material choices
- air filtration strategies
- fabric and furniture considerations
- practical steps to reduce microfiber shedding

This chapter gives you the starting point.

That book shows you how far you can take it.

The Takeaway

You don't just eat and drink your environment.

You breathe it.

> **Improve airflow.**
> **Filter what you can.**
> **Reduce what's being released.**

You won't eliminate exposure.

But you can lower it—every single day.

And when exposure is constant, daily reductions matter.

Chapter 15

Everyday Habits That Actually Matter

By now, you've seen the problem from every angle.

You understand:

- how nanoplastics form
- how they enter the body
- what they may do once inside
- and how to reduce exposure in your environment

At this point, it's easy to feel one of two things:

motivated—or overwhelmed.

This chapter is about making sure it's the first.

Because none of this works if it's too complicated to follow.

You Don't Need to Do Everything

One of the biggest mistakes people make is trying to change everything at once.

- replacing every item in their home
- eliminating all plastic overnight
- chasing perfection

That approach doesn't last.

And when it fails, people fall back to doing nothing.

This is not about perfection.
It's about direction.

You are not trying to eliminate exposure completely.

You are trying to **reduce it, consistently, over time.**

Focus on What Actually Moves the Needle

Not all actions are equal.

Some changes have a much bigger impact than others.

If you focus on the highest-impact behaviors, you get most of the benefit without the burden.

Start here:

1. Keep heat away from plastic
This is one of the most powerful and easiest changes.
Hot food and plastic should not mix.

2. Improve your air
Ventilation and filtration reduce what you breathe every day.

3. Filter your water (if possible)
Even partial improvement reduces daily intake.

4. Reduce dust exposure
Clean in a way that removes particles, not just moves them.

5. Be mindful of food contact
Packaging and processing matter more than most people realize.

These are not extreme changes.
They are high-leverage changes.

Build Habits, Not Projects

Big changes fail because they feel like projects.

Small changes succeed because they become habits.

Instead of:

- "I need to overhaul my entire kitchen"

Think:

- "I'll stop heating food in plastic starting today"

Instead of:

- “I need to eliminate all synthetic fabrics”

Think:

- “I’ll gradually replace the items I use most”

The goal is not intensity.
The goal is consistency.

When something becomes routine, it stops feeling like effort.

Reduce Friction

If a habit is difficult, it won’t last.

Make it easier:

- Keep glass containers accessible and ready to use
- Place a water filter where you naturally fill your glass
- Use a vacuum that’s easy to grab and use regularly
- Position air filters where you spend the most time

The easier it is, the more often it happens.

And frequency is what matters.

Avoid the Trap of Overwhelm

There is always more you *could* do.

That doesn’t mean you *should* do everything.

Information without structure leads to paralysis.

You don't need to solve the entire problem.
You just need to improve your part of it.

Every reduction counts.

Even if it feels small.

What Consistency Actually Does

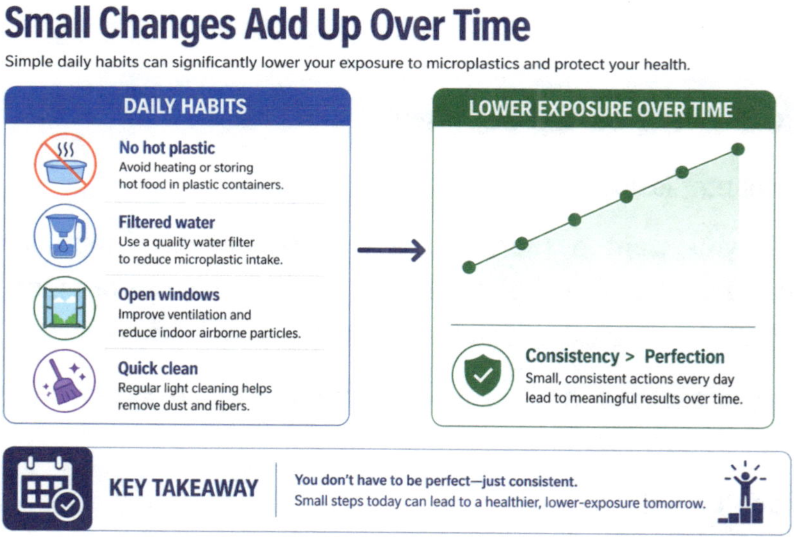

Nanoplastic exposure is not a single event.

It's cumulative.

That means small daily reductions matter more than occasional extreme efforts.

- One better choice repeated daily
- One habit maintained over time
- One exposure pathway reduced consistently

That's what changes your long-term burden.

Not a single action—but a pattern.

If You Want to Go Further

This chapter is about keeping things simple.

But if you want a more structured, step-by-step approach to reducing plastic exposure across your entire home environment, *The Plastic Free Home* expands on everything you've learned here.

It takes these ideas and turns them into a complete system you can follow room by room.

You don't need it to start.

But it's there if you want to go deeper.

The Takeaway

You don't need to do everything.

You need to do what matters—and keep doing it.

> **Start small.**
> **Focus on impact.**
> **Repeat it daily.**

That's how real change happens.

Not all at once—

but one habit at a time.

PART 6
The Bigger Picture

CHAPTER 16

Why This Problem Isn't Going Away

By now, you understand the personal side of the problem.

How nanoplastics form.
How they enter the body.
What they may do once they're inside.
And what you can do to reduce your own exposure.

But there's a bigger reality behind all of it:

> **This problem is not shrinking.**
> **It's growing.**

And it's growing for a simple reason:

We are still producing plastic at an enormous—and increasing—scale.

A World Built on Plastic

Plastic is everywhere because it solved real problems.

It's:

- lightweight
- durable
- inexpensive
- easy to manufacture

Those advantages made it indispensable.

Today, plastic is used in:

- packaging
- food systems
- healthcare
- construction
- clothing
- electronics

It is deeply embedded in modern life.

And that matters, because it means this isn't a niche issue.

It's systemic.

Production Is Still Rising

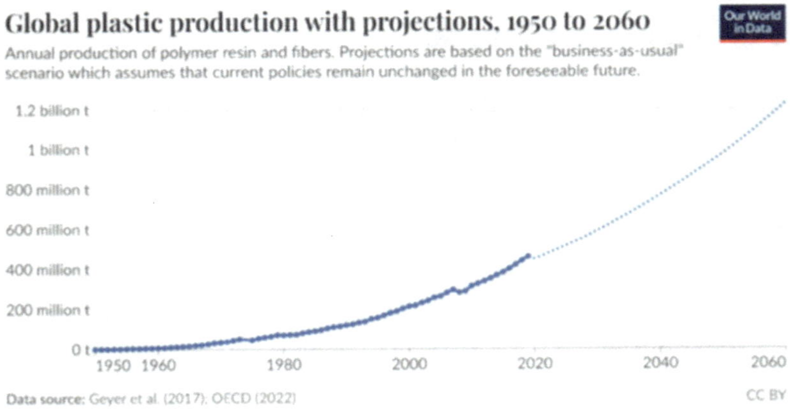

Source: Our World in Data (CC BY)

Global plastic production has increased dramatically over the past decades—and it continues to rise.

Each year:

- hundreds of millions of tons of plastic are produced
- a significant portion is designed for short-term use
- much of it becomes waste quickly

Even with increased awareness, the overall trend has not reversed.

> **More plastic produced means more plastic entering the environment.**

And over time, that leads to more fragmentation.

Plastic Doesn't Disappear—It Breaks Down

*Photo by **Aldward Castillo** on **Unsplash** (Unsplash License)*

One of the most important facts about plastic is also one of the simplest:

It does not go away.

Instead, it breaks apart.

Large plastic items:

- degrade through sunlight (UV exposure)
- break down through physical stress
- fragment through environmental processes

Over time, they become:

- microplastics
- and eventually, nanoplastics

This process happens:

- in oceans
- in soil
- in air
- and within built environments

Every piece of plastic ever created is on a path toward becoming smaller particles.

Persistence Means Accumulation

Because plastic does not biodegrade in the traditional sense, it accumulates.

Year after year, new plastic is added to what already exists.

That creates a compounding effect:

- more material in circulation
- more fragmentation over time
- more particles entering ecosystems and human environments

This is not a one-time pollution event.

It's an ongoing process.

And it builds.

Recycling Isn't a Complete Solution

Source: Our World in Data / OECD data (CC BY)

Recycling is often presented as the answer.

It helps—but it has limits.

Not all plastic is recyclable.
Not all recyclable plastic is actually recycled.
And recycling itself does not eliminate plastic—it often just reshapes it.

Even under ideal conditions:

- recycling rates remain relatively low globally
- plastic quality degrades with repeated processing
- and some recycled material still contributes to environmental shedding over time

Recycling slows the problem.
It does not stop it.

The Time Factor

Plastic production is not just large—it's continuous.

Every day:

- new plastic is manufactured
- new products enter circulation
- new waste is created

At the same time:

- older plastic is breaking down
- existing material is fragmenting further
- environmental reservoirs are expanding

This creates a timeline problem:

Even if production stopped today, fragmentation would continue for years.

But production hasn't stopped.

It's still increasing.

What This Means for the Future

The long-term implication is clear:

- more plastic in the system
- more breakdown into smaller particles
- more widespread environmental presence

Which leads to:

- continued human exposure
- ongoing scientific investigation
- and growing concern as more data emerges

This is not a problem that resolves on its own.

It requires change at a much larger scale.

What This Means for You

You cannot control global production.

But you can understand what it means.

> **This is not a temporary issue.**
> **It is a long-term environmental condition.**

That's why:

- reducing personal exposure matters

- awareness matters
- and informed choices matter

Not because they solve the entire problem—

but because they change your position within it.

The Takeaway

Plastic solved many problems.

But it created a new one.

And that problem doesn't disappear.

It fragments.
It spreads.
It accumulates.

> **As long as production continues,**
> **and as long as plastic persists,**
> **this issue remains.**

Understanding that is the first step toward addressing it—not just individually, but collectively.

CHAPTER 17

What Needs to Change (Policy & Industry)

By now, the pattern is clear.

Plastic doesn't disappear.
It breaks down.
It spreads.
And at the smallest scale, it begins interacting with biological systems in ways we are only beginning to understand.

Which leads to an uncomfortable truth:

> **This is not just a personal problem.**
> **It is a structural one.**

Individual choices matter.
But they operate inside a system that continues to produce—and distribute—plastic at massive scale.

If that system doesn't change, the problem doesn't either.

The Regulation Gap

Plastic has become essential to modern life.

But regulation hasn't kept up with its consequences.

Most current policies focus on:

- visible waste
- litter control

- recycling systems

What they often **don't fully address** is:

- fragmentation into micro- and nanoplastics
- long-term environmental persistence
- human exposure pathways

In other words:

> **We regulate plastic as waste.**
> **But we are only beginning to understand it as a long-term exposure issue.**

There is also inconsistency across regions:

- different countries apply different standards
- enforcement varies widely
- some materials are restricted in one place and unrestricted in another

This creates gaps that allow the problem to continue, even as awareness grows.

The Research Gap

The science is moving—but it's still catching up.

We now have:

- evidence of widespread environmental presence
- detection of micro- and nanoplastics in human tissues

- early indications of biological interaction

But major questions remain:

- How much exposure is too much?
- How do different particle sizes and chemistries behave in the body?
- What are the long-term health outcomes?

This is not unusual for emerging environmental issues.

But it creates a problem:

> **Policy often waits for certainty.**
> **Science produces certainty slowly.**

That delay matters.

Because during that time:

- exposure continues
- accumulation may increase
- and the system remains largely unchanged

More targeted research is needed:

- standardized measurement methods
- long-term human studies
- clearer understanding of nanoplastic behavior

Not to create fear—

but to replace uncertainty with clarity.

Industry's Role

Plastic production is not an accident.

It is driven by:

- demand
- cost efficiency
- manufacturing systems built over decades

That means industry is not just part of the problem.

It is central to the solution.

There are already signs of change:

- development of alternative materials
- redesign of packaging systems
- reduction of certain single-use plastics

But progress is uneven.

And in many cases:

- convenience still outweighs long-term impact
- short product lifecycles remain the norm
- materials are designed without full consideration of their end-of-life behavior

One of the key shifts needed is this:

Design with the full lifecycle in mind.

Not just:

- how a product is made
- or how it performs

But:

- how it breaks down
- where it ends up
- and what it becomes at the smallest scale

Because ultimately, every plastic product enters that trajectory.

Responsibility Across the System

It's easy to place responsibility in one place.

Blame consumers.
Blame industry.
Blame governments.

But the reality is more complex.

Each plays a role:

- **Consumers** influence demand
- **Industry** shapes what is available
- **Policy** defines the boundaries

Real change happens when those layers align.

When:

- better products are created
- better choices are available

- and better standards are enforced

That's when systems shift.

Why This Chapter Matters

It would be easy to end this conversation at personal behavior.

Use less plastic.
Filter your water.
Improve your indoor environment.

Those things matter.

But they exist within a larger system that continues to generate the problem.

Without systemic change, personal effort has limits.

That doesn't make it pointless.

It makes it incomplete.

The Direction Forward

The goal is not to eliminate plastic overnight.

That's unrealistic.

The goal is to move toward:

- smarter material use
- reduced unnecessary exposure
- better-designed systems

- and more informed decision-making at every level

This includes:

- improving regulation to reflect emerging science
- investing in research that answers real-world questions
- encouraging industry to design for long-term impact—not just short-term function
-

The Takeaway

Nanoplastics are not a distant issue.

They are the end result of a system we built.

> **If the system stays the same,**
> **the outcome stays the same.**

Understanding that doesn't create fear.

It creates clarity.

And clarity is what makes meaningful change possible.

CHAPTER 18

Final Thoughts: Living in a Plastic World

By now, you see the full picture.

Plastic is not just something we use.
It is something we live with.

It moves through our environment.
It breaks down into smaller and smaller pieces.
And at the nanoscale, it begins to intersect with the biology of the human body in ways we are only beginning to understand.

That reality can feel overwhelming.

But it doesn't have to be.

Clarity Over Fear

This book was not written to create fear.

It was written to create **clarity**.

Because fear without understanding leads to paralysis.
But understanding—real, grounded understanding—leads to better decisions.

You now know:

- where nanoplastics come from
- how they enter the body
- what emerging science suggests about their behavior

- and where the limits of current knowledge still exist

That last point matters.

> **There is still uncertainty.**
> **But there is no longer ignorance.**

You Are Not Powerless

It would be easy to walk away from this topic feeling like nothing can be done.

Plastic is everywhere.
Production continues.
Exposure is ongoing.

But that is only part of the story.

The other part is this:

> **You have influence over your environment.**

Not total control.
But meaningful influence.

You can:

- reduce exposure in your home
- make different choices with food and packaging
- improve your indoor air and water quality
- shift habits in ways that lower long-term burden

These are not extreme changes.

They are practical ones.

And over time, they matter.

Small Changes, Real Impact

There is a tendency to think in extremes:

- eliminate everything
- or change nothing

Neither approach works.

What does work is consistency.

Small, repeatable decisions:

- avoiding heating plastic
- choosing better materials when possible
- improving ventilation
- reducing dust and synthetic fibers

Each step may seem minor.

But together, they shift your exposure profile in a meaningful way.

> **The goal is not perfection.**
> **It is direction.**

Living With Reality, Not Avoiding It

Plastic is part of modern life.

It plays roles in:

- medicine
- safety
- infrastructure
- technology

This is not a problem that can be solved overnight—or by any one person.

So the goal is not to pretend plastic doesn't exist.

The goal is to **interact with it more intelligently**.

To understand:

- where it matters most
- where exposure can be reduced
- and where attention is worth your effort

That is what turns information into action.

The Bigger Picture Still Matters

Individual choices are powerful—but they are not the whole solution.

The larger system:

- how plastic is produced
- how it is used
- how it is managed

continues to shape the environment we all live in.

Change at that level takes time.

It involves:

- research
- policy
- industry shifts

But it starts with awareness.

Because systems do not change without pressure.
And pressure does not build without understanding.

What you now understand is part of that process.

What We Know—and What We Don't

The science of nanoplastics is still evolving.

We have:

- clear evidence of widespread presence
- growing evidence of human exposure
- early evidence of biological interaction

What we are still learning:

- long-term health effects
- accumulation patterns over time
- the full range of biological responses

That distinction matters.

Because it allows us to hold two truths at once:

> **There is enough evidence to take this seriously.**
> **And enough uncertainty to stay precise and honest.**

That balance is where credibility lives.

A Better Way to Think About It

You don't need to view this as a crisis you must solve alone.

A better way to think about it is this:

> **This is an environmental condition you now understand.**

And like any environmental condition, you can:

- adjust how you interact with it
- reduce unnecessary exposure
- and make informed decisions going forward

That is not helplessness.

That is awareness applied.

The Takeaway

Plastic is not going away tomorrow.

But your relationship with it can change today.

Not through fear.
Not through extreme measures.
But through informed, consistent action.

> **You don't need to eliminate plastic from your life. You need to understand where it matters—and act there.**

That is how you live in a plastic world—without being controlled by it.

And that is where real progress begins.

Biography

Robert Sudar is an author and advocate focused on the growing crisis of global plastic pollution. Driven by the alarming lack of public awareness regarding nano and micro plastics, he compiled years of global data to create a roadmap for individual and to advance systemic change. **Robert** lives in the United States and is committed to a future where the air, water, and soil are once again pristine, and the food supply is free of plastic contaminants.

Made in the USA
Monee, IL
05 November 2026

67811481R00101